Leckie×Leckie
Scotland's leading educational publishers

#1 FOR REVISION

National 5
DESIGN & MANUFACTURE
SUCCESS GUIDE

N5 DESIGN & MANUFACTURE *SUCCESS GUIDE*

Kirsty McDermid • Giorgio Giove
Francesco Giove

001/10102018

10 9 8 7 6 5 4 3 2 1

ISBN 9780008281946

Published by
Leckie & Leckie Ltd
An imprint of HarperCollins*Publishers*
Westerhill Road, Bishopbriggs, Glasgow, G64 2QT

T: 0844 576 8126 F: 0844 576 8131
leckieandleckie@harpercollins.co.uk
www.leckieandleckie.co.uk

Special thanks to
Project manager: Anna Clark
Layout: Ken Vail Graphic Design

Printed and bound by CPI Group (UK) Ltd, Croydon, CR0 4YY

A CIP Catalogue record for this book is available from the British Library.

Acknowledgements

We would like to thank the following for permission to reproduce photographs. Page numbers are followed, where necessary, by t (top), b (bottom), m (middle), l (left) or r (right).

P6 R.legosyn; p7 lightpoet; p10 Nejc Vesel; p15 Evgeny Karandaev; p31 donatas1205; p32 holbox; p36 Shell114; p39 Thinglass; p43 J and S Photography; p44t dmytro herasymeniuk; p44b Africa Studio; p45l Petr Malyshev; p45m Iakov Filimonov; p45r Bplanet; p48 yanugkelid; p49 Ruslan Semichev; p50t Africa Studio; p50b Neveshkin Nikolay; p51 steamroller_blues; p52l Anklav; p52r Alexey Boldin; p53tl Sashkin; p53tr art_of_sun; p53bl dean bertoncelj; p53br AlexMaster; p54 Milosz_M; p55 gpointstudio; p68 – Candidate A, Petr Malyshev (games controller), Krivosheev Vitaly (football), Room27 (bedroom), William Perugini (friends), DVARG (light bulb illustration), gillmar (energy-saving light bulb), gosphotodesign (light bulb & hand), Edler von Rabenstein (laptop), Marzanna Syncerz (boy), Zholobov Vadim (wood left), Voyagerix (wood middle), Superlime (wood right), Rmbssk, Mike Flippo (pizza), Marco Prati (anglepoise lamp), Peerawit (ceiling light); p68 – Candidate B, Kaziyeva-Dem'yanenko Svitlana (room), Alexey Boldin (blue watch), Robbi (shopping), tovovan (hand illustrations), Xtock Images (TV), PlusONE (modern interior), bikeriderlondon (white chair), Graham Taylor Photography (speakers), koya979 (table), Maksym Bondarchuk (lamp), BrAt82 (microphone), Monkey Business Images (teenager with phone), littleny (teenager with books), Ollyy (teenager on laptop), Tormod Rossavik (yellow lamp), dny3d (man with screwdriver), pterwort (extending lamp), Artit Fongfung (two red lamps), Darren Hubley (table lamp), Olga Selyutina (disco ball), dgbomb (Y light), Suwannakitja Chomraj (elephant lamp), Andrejs83 (lamp and book); p76 i3alda; p77 Dyson; p77 (biomimicry online resource) Stephen Mcsweeny; p78 jokerpro (toddler with pencils), Photographee.eu (woman on sofa), Peteri (wooden toy), Marco Govel (xylophone), chonrawit boonprakob (toy tools), ESOlex (toy building blocks), photosync (digger), aastock (vintage toy spinner), Cherry-Merry (children on gymnastic rings), Max Topchii (child with toy), Evgeny Karandaev (wooden train), conrado (boy playing with plane), Stepan Bormotov (ball); p87 Firma V; p86 Janna McDonald, Beath High School; p90 grafvision; p95t Joe Simon, Beath High School; p95b Nigel Rodger, Beath High School; p96 Jeffrey B. Banke; p102 Monkey Business Images; p103l nito; p103r Monkey Business Images; p104l ermess; p104ml Lichtmeister; p104mr McCarthy's PhotoWorks; p107 DJ Srki; p109t Palo_ok; p109ml Timmary; p109mr Vereshchagin Dmitry; p109bl Photoseeker; p109br pio3; p110tl LunaseeStudios; p110tm nikkytok; p110tr Kotomiti Okuma; p110ml Gyuszko-Photo; p110mr Oleksandr Chub; p110bl dyoma; p110br Sandra Cunningham; p111 (Scots Pine) American Spirit; p111 (Red Cedar) American Spirit; p111 (Ash) Igor Sokolov; p111 (Oak) PavelShynkarou; p111 (Beech) Evgeny Karandaev; p111 (Mahogany) Angel_Vasilev77; p111 (MDF) leungchopan; p111 (Plywood) Hassel Sinar; p111 (Chipboard) Pawel Gr; p111 (Hardboard) Voyagerix; p111 (Manufactured Board) hasan eroglu; p113tl Dandesign86; p113tr titov dmitriy; p113ml Maxx-Studio; p113mm Cathleen A Clapper; p113mr zhu difeng; p113bl Garsya; p113br Innershadows Photography; p115 (Injection Moulding) MidoSemsem; p115 (Vacuum Forming) Timof; p115 (Die Casting) Digital Genetics; p115 (Rotational Moulding) pavla; p115 (Extrusion) Dmitry3D and Oleksiy Mark; p115 (Turning) ozguroral; p117t KPG Payless2; p117m Huguette Roe; p117b Pedro Miguel Sousa; p120b Lersak supamatra; p121t Photobac; p121b John Kasawa; p122tl lightwavemedia; p122tr RTimages; p122b Axel Bueckert; p123t Visaro; p123b Allard One; p124t tale; p124b leungchopan; p125tl Nenov Brothers; p125tm nikkytok; p125tr Peter Turner Photography; p125bl indigolotos; p125br brulove

Whilst every effort has been made to trace the copyright holders, in cases where this has been unsuccessful, or if any have inadvertently been overlooked, the Publishers would gladly receive any information enabling them to rectify any error or omission at the first opportunity.

Kirsty, Giorgio and Francesco would like to thank:

Nathan, Annez and Lisa for your support over the past months, despite being abandoned in favour of this book

Dick for going beyond the call of duty in your feedback and guidance

Janna, Nigel and Joe for allowing us to use your work

Ben and Emily for letting us use your photograph

And finally Fiona and Anna for your support in helping us pull this together.

ebook

To access the ebook version of this Success Guide visit

www.collins.co.uk/ebooks

and follow the step-by-step instructions.

Contents

Contents

Course Rationale

In Product Design, Architecture, Engineering and many other design-based disciplines, creativity and problem-solving skills are fundamental.

The National 5 Design and Manufacture Course reflects these industries and provides a broad practical introduction to design, materials and manufacturing processes. You will be able to explore your creative abilities, as well as develop problem-solving skills in a range of design contexts.

The Course also brings together science, mathematics and technology to partner design and creativity.

The emphasis of the N5 Design and Manufacture Course is on developing a range of skills. From generating ideas and presenting a design proposal, to making and testing a prototype, the Course is very much a practical experience.

The aims of the Course, as detailed by the SQA, are to enable you to develop:

- skills in design and manufacturing of models, prototypes and products
- knowledge and understanding of manufacturing processes and materials
- an understanding of the impact of design and manufacturing technologies on our environment and society.

Core Transferable Skills

The National 5 Design and Manufacture Course will provide you with skills that will help you to learn, live and work effectively in a society where technology is always improving and advancing.

Many of the skills you learn in National 5 Design and Manufacture will help you in other subjects in your school work, as well as helping prepare you for life after you leave school. You will learn to apply creative thinking to solve complex problems, work independently and effectively, develop skills to communicate visually and develop your confidence as you apply new skills in a context.

Course Overview

To gain both of the N5 Area of Study awards, you must pass all of the Assessment Standards within each of the Areas of Study. To achieve the N5 Course award you will need to have covered all of the skills and knowledge that are in the Areas of Study to ensure you are prepared to sit the Course Assessment.

Area of Study 1 – Design

This Area of Study focuses on developing your design knowledge, skills and creativity to enable you to design products that are suitable for manufacture. You will learn about the design process and factors that influence design, as well as techniques for creating, communicating, researching and evaluating.

You will develop research, problem-solving and evaluating skills as you work through design projects.

During this Area of Study you will also learn about different approaches to communicating that are appropriate to the stage in the design process. These approaches include both graphic and modelling techniques.

Area of Study 2 – Materials and Manufacture

This Area of Study focuses on developing your knowledge of materials, manufacturing processes and practical skills, which are all required for manufacture. You will produce a detailed plan for manufacture, as well as demonstrate your practical skills by creating a range of models and prototypes.

This Area of Study focuses on developing your knowledge and understanding of a range of materials and processes that are used in workshop and commercial contexts.

For workshop-based tasks, you will be required to select appropriate materials and justify your choices, produce and evaluate a detailed plan for manufacture, and demonstrate your practical skills through the manufacture of a range of products.

During this Area of Study you will also learn about materials and processes that are used in the commercial production of familiar products, and the impact the design and manufacture of these products has on the environment and society.

Where do you like to study? This design student has found an unusual study space, but it is free from distractions and she can focus completely!

Course Assessment

How You Will Be Assessed

There are three elements to the Course Assessment: Assignment Design, Assignment Practical and the Question Paper.

Your Course Assessment will be graded A–D depending on how well you do.

There are 180 marks available in total.

30.6% of the marks are allocated to Assignment Design

25% of the marks are allocated to Assignment Practical

44.4% of the marks are allocated to the Question Paper

Assignment Design

You will have to apply your design skills and knowledge from the Areas of Study to produce a creative solution to one of the design tasks that are set by the SQA.

There are 55 marks available for Assignment Design, 30.6% of the overall marks.

Assignment Practical

You will have to apply your practical skills gained in the Areas of Study to manufacture your proposal from Assignment Design. You will be required to carry out an evaluation of the manufactured prototype.

There are 45 marks available for Assignment Practical, 25% of the overall marks.

The Question Paper

The question paper is worth 80 marks, 44.4% of the overall marks.

You need to apply your knowledge gained in the Areas of Study to answer questions about design, materials, manufacturing in the workshop and commercial manufacture.

37.5% of the question paper will test K&U of design

37.5% will test workshop-based manufacture

25% of the question paper will assess K&U of commercial manufacture

You will be expected to **explain**, **describe** and **justify** to demonstrate your understanding. The paper is split into two sections. Section 1 has six or seven questions, is worth 60 marks and assesses knowledge of design and workshop-based manufacture. Section 2 has four or five questions, is worth 20 marks and assesses knowledge of commercial manufacture.

Preparing well will allow you to approach your exam with confidence.

How to Use This Success Guide

The Success Guide has been written to give you practical support throughout your Coursework, during the Assignment and to prepare you for the Question Paper.

You should use this success guide when:

- working on the Areas of Study in class
- revising for tests
- revising for your prelim or the Question Paper
- preparing for and carrying out the Assignment.

Throughout the Success Guide you will find good and not-so-good examples of work to help build your understanding of what is required to pass the Standards and to acheive the different elements of the Course Assessment. The examples will support your understanding as you work through the Areas of Study in class. The book also has two sections dedicated to helping you reach your potential in the Course Assessment.

Success Guide Features

The Success Guide is packed with hints and tips, skill builders and questions to help you along the way.

GOT IT? ☐ ☐ ☐

You will see the 'Got it?' boxes at the top of the pages. This traffic-light system enables you to record your understanding and can be used at any time throughout the Course and for your revision.

☐ Tick the red box when you have little or no understanding of the topics in this area. You should focus on improving your knowledge and understanding in these areas during future revision.

☐ Tick the amber box when you have some understanding but there are areas that you still need to revise or improve.

☐ Tick the green box when you are confident you understand and can demonstrate your knowledge and skill in this area.

Question Paper/ Assignment Tip

These tips will relate directly to the Course Assessment Assignment or Question Paper.

Advice

You will find Advice boxes throughout the Success Guide.
These are designed to draw your attention to specific areas within the topics.

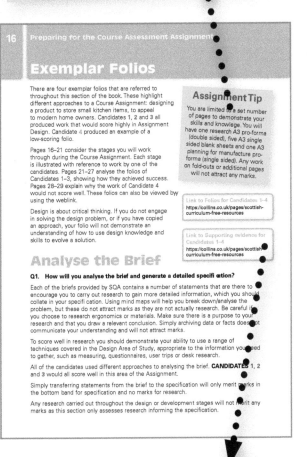

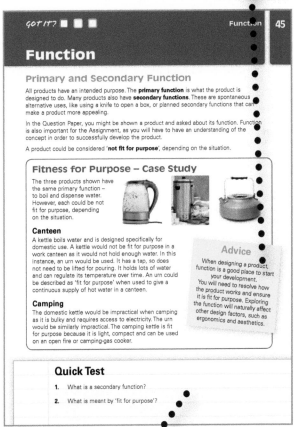

You will find weblinks that direct you to resources to help with your revision.

Quick Test

1. You will find questions at the bottom of the pages. These test your knowledge and understanding, and allow you to summarise at the end of each section.

   Answers to these questions can be found at the back of the book (pages 137–139).

Revising

Using the Glossary

You will come across new words as you progress through the National 5 Design and Manufacture Course. You will find a glossary on pages 130–133 that will help you keep on top of your new vocabulary.

A glossary is like a dictionary; words and their meanings are listed alphabetically. Refer to it when you come across a word in this Success Guide that you don't know.

Planning and Tracking Your Revision

This Success Guide aims to help you study effectively. You will have other subjects to revise, leading up to tests, prelims and the final exams. It is important to plan your studying and use your time wisely to ensure you are well prepared for all of your exams.

Here are some tips to help you plan your revision:

- Organise and plan your revision as early as possible. 'Cramming' last-minute revision the night before an exam is an ineffective way to revise.

- Break down your revision into manageable chunks.

- Make notes and then commit enough time to revise them. Your priorities may change from week to week so forward planning is important.

- The time leading up to exams can be stressful. It is important to find a balance in your life and continue to do things you enjoy. This time out will ensure you are better focused when you are revising.

- You can become mentally fatigued if you try to study for too long at a time. Study in blocks of time, with regular short breaks between. Do not try to study for any longer than you can concentrate.

You can try this effective method to plan and record your revision. It takes a little bit of time to prepare but will be hugely beneficial leading up to your final exam.

Firstly, you will need to make three columns and colour code them using the traffic-light method. Use the 'checklists' from this Success Guide to create lists to help you prioritise areas for revision, as shown below.

Topics I know little about	Topics I understand a bit	Topics I am confident about
Ergonomic influences in design	Idea-generation techniques	Common hand tools
Modelling techniques in the design process	Working drawings	Knowledge: plastics, properties and uses
Metals, their properties and uses	Commercial manufacture	Evaluation strategies
Assembly - woodwork joining methods		

Once you have made a list like this, you can use a Gantt chart to plan how much time you will spend revising each topic. A Gantt chart shows the list of tasks down the left-hand column and bars next to these tasks indicate how much time will be dedicated to each specific task.

Topic	Week 1	Week 2	Week 3	Week 4
Ergonomic influences in design	▨▨▨			
Modelling techniques in the design process.		▨▨▨		
Metals, their properties and uses.			▨▨▨	
Assembly - woodwork joining methods				▨▨▨▨
Idea-generation techniques		▨▨▨		
Commercial manufacture				▨▨▨

It is important to be realistic about the time you will commit to your revision. It is better to aim for less time and then do more. You will be disappointed if you set unrealistic targets and don't achieve them.

Area of Study 1: Design – Revision Checklist

Design Process

By confidently applying skill, knowledge and understanding, I can:

GOT IT?

- carry out **research** using a range of techniques such as search engines, measuring and recording, asking questions, surveys and using data ☐ ☐ ☐

- apply a range of **idea-generation techniques** such as morphological analysis, brainstorming, technology transfer, analogy, lateral thinking, six-hat thinking and SCAMPER ☐ ☐ ☐

- **develop** by **exploring** and **refining** ideas, and justify and record decisions made ☐ ☐ ☐

- **evaluate** design work by linking back to a **specification**. ☐ ☐ ☐

Design Factors

I can apply my knowledge and understanding of the following design factors to develop a product:

- **function** (primary and secondary functions, fitness for purpose) ☐ ☐ ☐

- **performance** (ease of maintenance, strength and durability, ease of use) ☐ ☐ ☐

- **market** (consumer demands, social expectations, branding, introduction of new products, market segments, marketing mix, needs, wants, technology push, market pull) ☐ ☐ ☐

- **aesthetics** (shape, proportion, size, colour, contrast, harmony) ☐ ☐ ☐

- **ergonomics** (establishing critical sizes, basic understanding of how humans interact with products, anthropometrics). ☐ ☐ ☐

Communication Techniques

Using learned techniques, knowledge and understanding, I can apply skillfully:

- a range of **graphic techniques** (working drawings, annotated sketches and drawings, perspective sketches, illustration and presentation techniques, scale and proportion, simple orthographic drawings) ☐ ☐ ☐

- a range of **modelling techniques** (scale models, use of appropriate modelling materials or CAD models). ☐ ☐ ☐

Materials and Manufacture – Revision Checklist

Planning for Manufacture

By confidently applying skill, knowledge and understanding, I can:

GOT IT?

- plan for practical tasks, create a cutting list, produce dimensioned drawings or sketches, plan a logical sequence for manufacture including steps, tools and equipment, select appropriate finishes. ☐ ☐ ☐

Tools, Materials and Processes

By confidently applying skill, knowledge and understanding, I can:

- use a range of common **workshop tools** to measure, mark out, cut, shape and form materials ☐ ☐ ☐

- use appropriate and acceptable **tools or equipment** to hold, clamp and restrain materials ☐ ☐ ☐

- use a range of common and acceptable **machine tools** to sand, shape and drill materials ☐ ☐ ☐

- produce a range of **fixing and joining techniques** by jointing/ joining woods, metals and plastics, including thermal joining and adhesive bonding ☐ ☐ ☐

- use a range of **metal-manufacturing processes** to cut, shear, notch, parallel and step turn, taper turn, drill, knurl, file, fold and bend metals ☐ ☐ ☐

- use a range of **plastic-manufacturing processes** to cut, drill, file, bend, twist and mould plastics. ☐ ☐ ☐

Properties of Common Materials

By confidently applying knowledge and understanding, I can:

- **select materials** such as softwoods, hardwoods, manufactured boards, ferrous metals, non-ferrous metals, thermoplastics and thermoset plastics that are appropriate for a product solution. ☐ ☐ ☐

Health and Safety

I can, at all times, demonstrate and adhere to **safe working practices** when undertaking manufacturing tasks in a workshop environment. ☐ ☐ ☐

Design Stage

The SQA will provide a selection of design briefs for you to choose from. All of the tasks will assess your ability to apply the design process, develop and communicate a solution to the problem and manufacture a prototype of the design. Your design must allow you to demonstrate skill in manufacturing.

The following questions should help you identify any areas that you may need to revise to improve your understanding. Further information and guidance can be found on the pages indicated below:

Analyse the Brief – Page 16

Explore Ideas – Page 17

Understanding Design Issues – Page 18

Understanding Materials and Manufacturing – Page 18

Ongoing Review of Ideas – Page 19

Communication – Page 20

Research and Time – Page 21

☐ ☐ ☐ Q1. How will you analyse the brief and generate a detailed specification?

☐ ☐ ☐ Q2. How will you generate your ideas?

☐ ☐ ☐ Q3. How will you explore ideas?

☐ ☐ ☐ Q4. How will you refine your ideas?

☐ ☐ ☐ Q5. How will you demonstrate the application of your knowledge and understanding of appropriate design factors?

☐ ☐ ☐ Q6. How will you demonstrate the application of your knowledge and understanding of appropriate materials and manufacturing?

☐ ☐ ☐ Q7. How will you record your decisions throughout the design process?

☐ ☐ ☐ Q8. How will your use of graphic techniques change as you progress through the folio and why?

☐ ☐ ☐ Q9. How will your use of modelling change as you progress through the folio and why?

☐ ☐ ☐ Q10. How will you plan the manufacture of your product?

☐ ☐ ☐ Q11. How will you manage your time during different stages of the design process?

Manufacturing Stage

Once you have successfully refined your idea, you will need to manufacture it in a given timescale. During the manufacturing part of the Assignment, you must demonstrate a high level of skill in marking out, cutting, shaping and/or forming, using hand and machine tools, assembling and finishing the prototype.

Read through the questions below to identify any areas of the manufacturing stage of the Assignment you may be unsure of. Further information and guidance on the skills you need to demonstrate can be found on pages 30–33.

☐ ☐ ☐   Q1.   What can you manufacture for the final product?

☐ ☐ ☐   Q2.   What will you do to ensure you achieve good marks for marking out?

☐ ☐ ☐   Q3.   How will the accuracy of your work be assessed?

☐ ☐ ☐   Q4.   What will you do to ensure you achieve good marks for using hand and machine tools?

☐ ☐ ☐   Q5.   What will you do to achieve good marks for the assembly of your product?

☐ ☐ ☐   Q6.   What will you do to ensure you achieve good marks for the quality of finish on your product?

☐ ☐ ☐   Q7.   How will you evaluate your manufactured proposal?

☐ ☐ ☐   Q8.   What supporting evidence can be submitted when your work is being sent to the SQA?

Assignment Tip

You must ensure that the product you design enables you to demonstrate your practical skills at an appropriate level. Overly simple products will not allow you to access the full range of marks available in the Manufacturing stage.

Exemplar Folios

There are four exemplar folios that are referred to throughout this section of the book. These highlight different approaches to a Course Assignment: designing a product to store small kitchen items, to appeal to modern home owners. Candidates **1**, **2** and **3** all produced work that would score highly in Assignment Design. Candidate **4** produced an example of a low-scoring folio.

Pages 16–21 consider the stages you will work through during the Course Assignment. Each stage is illustrated with reference to work by one of the candidates. Pages 21–27 analyse the folios of Candidates 1–3, showing how they achieved success. Pages 28–29 explain why the work of Candidate 4 would not score well. These folios can also be viewed by using the weblink.

Design is about critical thinking. If you do not engage in solving the design problem, or if you have copied an approach, your folio will not demonstrate an understanding of how to use design knowledge and skills to evolve a solution.

Assignment Tip

You are limited to a set number of pages to demonstrate your skills and knowlege. You will have one research A3 pro-forma (double sided), five A3 single sided blank sheets and one A3 planning for manufacture pro-forma (single sided). Any work on fold-outs or additional pages will not attract any marks.

Link to Folios for Candidates 1–4.
https://collins.co.uk/pages/scottish-curriculum-free-resources

Link to Supporting evidence for Candidates 1–4
https://collins.co.uk/pages/scottish-curriculum-free-resources

Analyse the Brief

Q1.   How will you analyse the brief and generate a detailed specification?

Each of the briefs provided by SQA contains a number of statements that are there to encourage you to carry out research to gain more detailed information, which you should collate in your specification. Using mind maps will help you break down/analyse the problem, but these do not attract marks as they are not actually research. Be careful if you choose to research ergonomics or materials. Make sure there is a purpose to your research and that you draw a relevant conclusion. Simply archiving data or facts does not communicate your understanding and will not attract marks.

To score well in research you should demonstrate your ability to use a range of techniques covered in the Design Area of Study, appropriate to the information you need to gather, such as measuring, questionnaires, user trips or desk research.

All of the candidates used different approaches to analysing the brief. **CANDIDATES** 1, 2 and 3 would all score well in this area of the Assignment.

Simply transferring statements from the brief to the specification will only merit marks in the bottom band for specification and no marks for research.

Any research carried out throughout the design or development stages will not merit any marks as this section only assesses research informing the specification.

Generate and Explore Ideas

Having a detailed specification is an essential part of your Assignment. It is crucial to have further detail to ensure when you start designing, you have things to explore and refine. Referring to the points in a good specification when making design decisions will also help you demonstrate your application of design knowledge.

Candidate 3 carried out a questionnaire to identify the most popular items to be stored and preferred materials. They also measured the popular items.

Q2. How will you generate your ideas?

There is no set method or predetermined number of ideas that you need to put forward to pass this part of the Assignment. Applying idea-generation techniques, such as those learned in the Design Area of Study, will enhance your ability to easily generate a range of diverse and creative ideas.

All of the candidates used a different approach to analysing the brief. **CANDIDATES 1, 2** and **3** would all score well in this area of the Assignment.

Remember, the point of initial ideas is to quickly generate something that has the potential to be a viable solution. Do not waste unnecessary time at this stage on presentation. A designer wouldn't!

Use annotation and a range of graphic and/ or modeling techniques to communicate the design details, function of the product and other significant aspects of the specification.

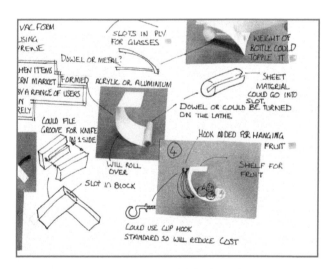

Candidate 3 chose to generate initial ideas using models. They added useful comments and additional sketches to clarify aspects of each idea.

Understanding Design Issues, Materials and Manufacturing

Q3. How will you explore ideas?

You should think about the design issues that you still need to resolve to work through a variety of alternative solutions. Creating a *development plan* (page 80) is a simple and effective way to ensure you solve all of the design problems that are *unique* to your project. Some ideas might seem silly at first, but every one is worth exploring. You will be credited for explaining why some ideas would not be successful.

CANDIDATES 1, **2** and **3** would also score well in this area of the Assignment. They have explored areas including function, ease of use, aesthetics and assembly. Their exploration is meaningful and has allowed them to develop an improved solution.

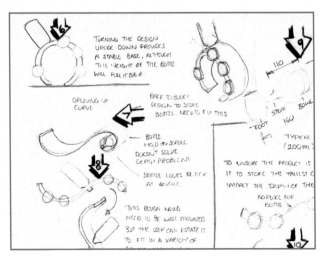

In this extract, Candidate 2 explores aesthetics, function and safety. Arrows are used to clarify their path.

Q4. How will you refine your ideas?

Refining ideas is about making decisions, removing unwanted aspects from the design and clarifying the details to permit manufacture. Modeling is a useful tool to refine ideas and help you make decisions to ensure your solution meets the requirements of the specification. You will probably need to finalise details of the aesthetics, function, ergonomics, materials selection, dimensions and assembly details.

CANDIDATES 1, **2** and **3** have all developed their solution using the specification. Their solutions meet the requirements of the brief and have sufficient detail to permit manufacture.

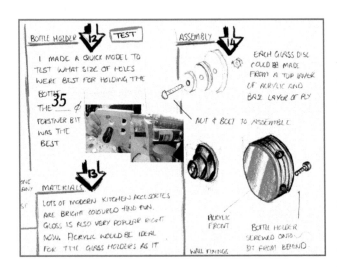

In this extract, Candidate 2 is considering the sizes required for the product to function and how the product could be manufactured.

Ongoing Review of Ideas

Q5. **How will you demonstrate the application of your knowledge and understanding of appropriate design factors?**

You should have a good understanding of design factors. Exploring aspects of your design will give you opportunities to apply your knowledge of design. You can demonstrate your knowledge by recording your decisions. Your knowledge will help you develop the product in a meaningful way, help you make good decisions and evolve your design to meet the specification.

CANDIDATES 1, 2 and 3 would all score well in this area of the Assignment. They have all used their design knowledge to inform the development of their ideas.

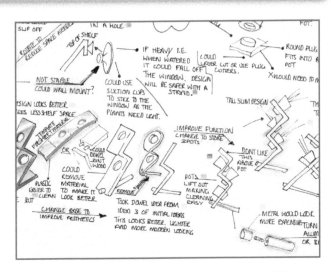

In this extract, Candidate 3 has used different coloured marks, indicating the area of the specification and design factor they are exploring. No marks are awarded for colour coding; however, it allowed the candidate to see they had applied knowledge of a range of design factors.

Q6. **How will you demonstrate the application of your knowledge and understanding of appropriate materials and manufacturing?**

Throughout the process, you have opportunities to draw on your knowledge of materials and manufacturing. You should make informed choices about materials and finishes. You should also demonstrate your understanding of the tools and processes that could be used to manufacture and assemble the solution. You must show this through annotation throughout the development of the solution. Knowledge of materials and processes contained in the planning for manufacture pro forma will not be awarded marks for this section.

CANDIDATES 1, 2 and 3 have all communicated a good understanding of materials and manufacturing throughout the development and refinement of their design.

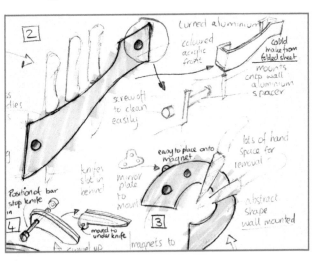

In this extract, Candidate 1 has considered alternative methods and materials for the shelf. They have demonstrated a good understanding of materials and processes relevant to their design.

Communication

Q7. How will you record your decisions throughout the design process?

Your folio of work should be easy for anyone, but especially the assessor, to follow. Any decisions you have made should be clear, as should your reasons for making them. You will need to make many decisions including:

- which ideas to reject and which to take forward

- how the product will look

- the selection and suitability of materials

- sizes and dimensions of the product

- how the product will function

- how the product will be assembled and finished.

The candidates have all used different methods to record their decisions. **CANDIDATES** **1**, **2** and **3** would all score well in this area. Candidate **4** would not. Not exploring the design has meant few decisions could be made.

Q8. How will your use of graphic techniques change as you progress through the folio and why?

This stage is marked by looking at all pages in your folio. As a general rule, initial ideas should be quick, with development sketches improving in accuracy and technical detail. Although initial ideas should be quick, they should still communicate basic details of the design, such as the thickness of material.

You are expected to demonstrate a range of graphic techniques that are appropriate to the stage in the design process. This includes pictorial views, working drawings and assembly details.

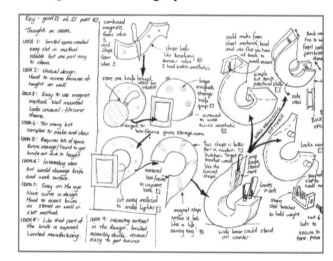

In this extract, Candidate 1 has used a simple traffic light system to communicate good and bad aspects of their ideas as they work through the process. The traffic lighting doesn't attract marks, but has helped the candidate visualise strengths and weaknesses, which they have explained in their annotations.

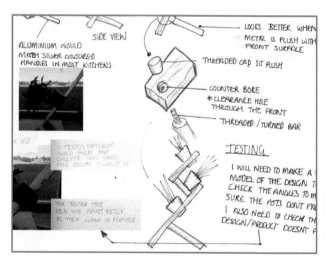

In this extract, Candidate 3 has used 2D and 3D graphics. They have produced an exploded detail to communicate how the product might assemble. Note, rendering can be time consuming and is not necessary to achieve full marks in using graphics.

Research and Time

Computer generated graphics can be used, as well as manual rendering techniques, to demonstrate a wider range of skills.

CANDIDATES 1, 2 and 3 have all used a range of graphic techniques throughout their design work. Candidate 4's graphics are limited and fail to communicate more detail in the later stages of their work.

Q9. How will your use of modelling change as you progress through the folio and why?

Modelling can be used at any stage: to generate ideas, develop and refine a solution or to communicate aspects of your design proposal. Models of initial ideas are likely to be generated quickly and be rough in nature. In development, models are invaluable for problem solving. They can be made to test and evaluate aspects of the design, such as function or ergonomics. Models should be appropriate to the stage in the process and must provide additional information that you could have not already shown in a sketch.

CANDIDATES 1, 2 and 3 have used models purposefully to solve problems in their design task. Candidate 4's model serves little purpose. It is quick model, not to scale, that simply gives a 3D view.

The extract from candidate 3's design work (above) shows quick card models that were used to test the size of hole and angle required to hold the plant pots.

Q10. How will you plan the manufacture of your product?

Your plan for manufacture must be completed on the pro forma supplied by SQA. You should have a cutting list detailing the sizes and materials of the component parts, a dimensioned drawing or sketch, and a logical sequence of operations that details the tools and equipment required at each stage.

It is important that you do not change the size of the boxes on the sheet as any change to the layout will result in you not receiving any marks for this section. Any graphics or models used at this stage will be credited in the appropriate sections; however, knowledge of design and/or materials and manufacture will not attract marks in this section.

Q11. How will you manage your time during different stages of the design process?

Your time for the Assessment is limited. Ideas should be generated quickly. Spend most of your time on development, as this is where the meaningful work takes place (and the marks are allocated). Here you will demonstrate your ability to apply graphic and modelling skills, and knowledge of design issues, and materials and manufacturing. You will use this knowledge to inform decisions as you work through the design process.

Some activities waste time and do not attract marks; drawing and colouring headings and borders for example. So, *do not* do it. Rendering every sketch is also unnecessary. A few good examples will demonstrate that you have the skill without wasting time.

It is clear from the work of **CANDIDATES** 1, 2 and 3 that they have invested considerable time in the development stage and would score highly in this area. CANDIDATE 4 has wasted time on a number of activities that do not attract marks. Producing evaluation tables not only takes up a lot of space, but can also be time consuming. On their own, they will not attract any marks as they do not demonstrate understanding or reasons for decisions.

Candidate 1

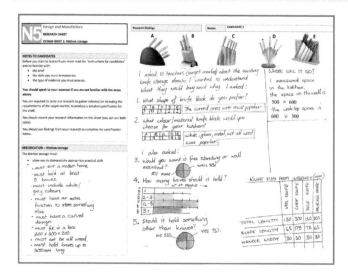

Q1 Analysing the brief:
CANDIDATE 1 has used a a range of methods effectively to gather valid information about their task. They used a questionnaire to find preferences in knife storage, secondary function and direction on aesthetics. They have also gathered measurements and incorporated their findings along with points from the brief, into a detailed specification.

Q2 Generating ideas: They have used *lifestyle boards* and *taking your pencil for a walk* to generate their initial ideas. All of the candidate's ideas are for products to store knives. However they are all very diverse in appearance and in how the user interacts with them. Creativity is evident in the unusual aspects of their ideas.

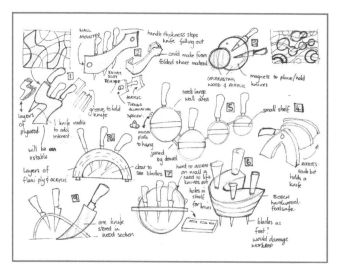

Q3 Exploring ideas: They have carried out meaningful exploration towards a solution, considering the aesthetics of the product, how the knives are held, additional storage space, materials and assembly.

Q4 Refining ideas: They explored aspects of the design and recorded clear decisions throughout the development, ensuring that they were moving towards the best solution. Modelling helped refine sizes, assembly and ease of use.

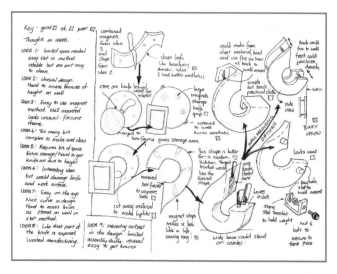

Q5 Using design factors: They have applied their knowledge and understanding of design factors, resulting in *meaningful* development of the product. It is clear from their evaluation of initial ideas that they understand the significance of ergonomics, function and aesthetics for their product. Annotations throughout the exploration support this.

Q6 Materials and manufacture: In their annotation they have clearly considered a range of suitable materials and processes that could be used in the manufacture of the product.

Q7 Recording decisions: They have made clear decisions by evaluating their ideas using good annotations throughout their development.

Q8 Using graphic techniques: Initial ideas have been sketched quickly and they have attempted some isometric sketches. During development they have shown assembly details, a detailed working drawing and evidence of some rendering.

Q9 Using models: A model was used effectively during the development stage to demonstrate the function of components. This was appropriate as it was clearer, easier and less time consuming than creating a series of complex sketches.

Q10 Planning manufacture: They have created a detailed cutting list and three orthographic views that have sufficient dimension and information relating to assembly. The candidate also has a detailed sequence that highlights the equipment needed at each stage.

Q11 Time management: It is clear from the work of **CANDIDATE 1** that they have invested considerable time in the development stage and would score highly in this area.

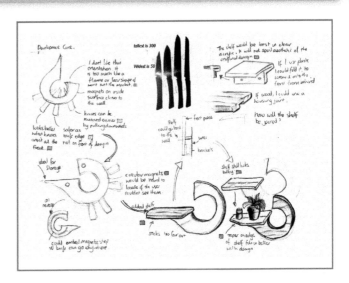

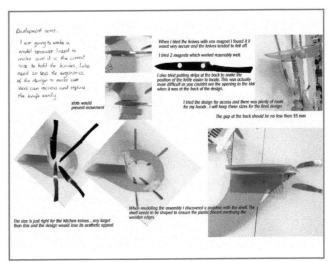

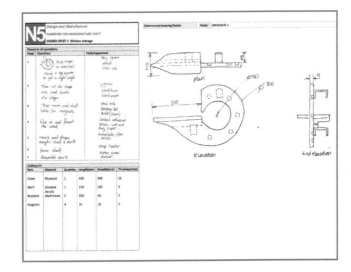

Candidate 2

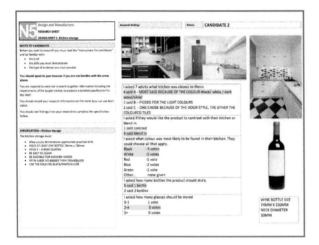

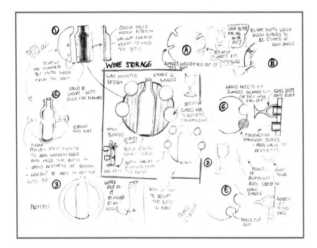

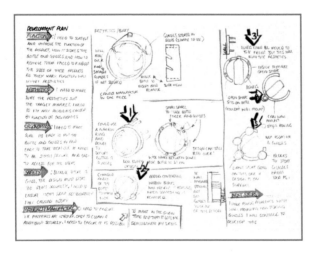

Q1 Analysing the brief: CANDIDATE 2 has used a a range of methods effectively to gather valid information about their task. They used a questionnaire to identify the number of bottles and glasses and direction on aesthetics. They have also gathered measurements of the bottle and created a detailed specification. The research on glass sizes carried out during development is not awarded marks in this section as it doesn't inform the specification.

Q2 Generating ideas: They have generated one initial idea. Creativity is evident in the unusual form and function of the design. Although there is only one concept, the candidate has shown diversity by giving alternatives for different aspects of the design. Ideas are clearly relevant and well annotated. This candidate has demonstrated idea generation in their idea development work.

Q3 Exploring ideas: They have shown exploration within the initial ideas page. They have also explored a variety of aspects throughout the development.

Q4 Refining ideas: They have added significant detail throughout the development. They used models to refine the idea. The CAD model illustrates the assembly method and specifies the sizes. Models were used to solve the ergonomic problems, and finalise sizes and manufacturing details. The development is *logical* and *justified*.

Q5 Using design factors: They have used a development plan, demonstrating their understanding of appropriate design factors. Additional research and modelling is used effectively with the clear purpose of refining a range of design issues.

Q6 Materials and manufacture: This candidate has justified materials and processes within the development of the proposal.

Q7 Recording decisions: This candidate used a development plan. They have clearly detailed their intentions and explained the decisions they have made.

Q8 Using graphic techniques: They have used simple 2D sketches to communicate their initial ideas. They have used isometric and oblique sketches to communicate their development. A CAD graphic has been produced to show assembly details. They have also demonstrated rendering in their design proposal.

Q9 Using models: This candidate has created a simple model to help them resolve size issues and identify ergonomic problems. They also created a test part, drilling different holes to test the stability of the bottle.

Q10 Planning manufacture: They have produced a detailed CAD orthographic with sufficient dimensions and a detailed sequence that summarises the tools and equipment needed at each stage. The candidate has not completed a cutting list; however, they would still score reasonably well for the detail in the rest of the plan.

Q11 Time management: CANDIDATE 2 has invested considerable time in the development stage and would score highly in this area.

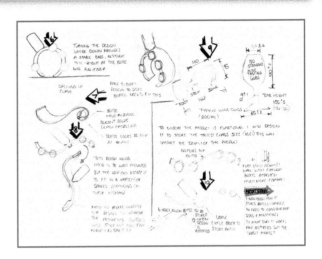

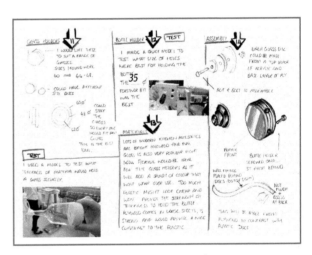

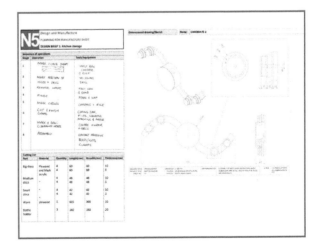

Candidate 3

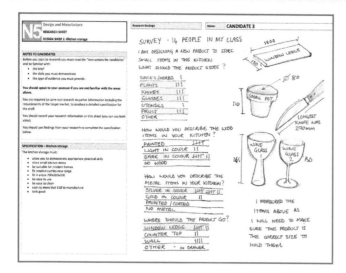

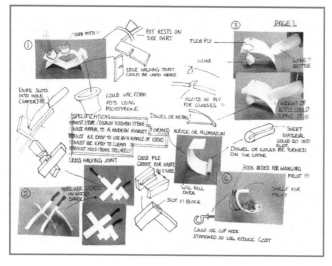

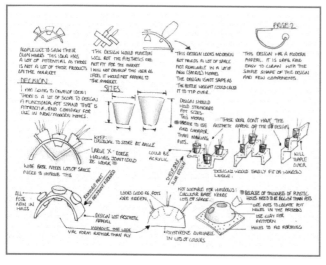

Q1 Analysing the brief:
CANDIDATE 3 has measured a range of items and used questionnaires effectively to gather valid information about the aesthetics and items to be stored for their task. The specification is detailed; however, they didn't make a decision on what the product was to hold at this stage. Research of pot sizes later in the folio doesn't attract marks as it didn't inform the specification.

Q2 Generating ideas: They have used *morphological modelling* to generate creative ideas that solve a variety of kitchen storage problems. The candidate demonstrates diversity in the function of the ideas, choice of materials and aesthetics. Annotations provide additional information to clarify aspects of the ideas.

Q3 Exploring ideas: The candidate has applied some understanding of design issues, and materials and manufacturing, considering alternative ways to manufacture their product.

Q4 Refining ideas: They have used a combination of models and sketching to finalise different aspects of their design. Each method was appropriate for the purpose. There is sufficient detail available to enable the product to be manufactured. *Informed decisions* have been made from the exploration to ensure the design has been developed to meet all areas of the specification.

Q5 Using design factors: They have applied some understanding of design factors within the development of their solution. A clear understanding is demonstrated in the detailed evaluation of their solution.

Q6 Materials and manufacture: They have looked at different options for manufacture, justifying the most suitable methods for their solution. They have clearly referenced materials throughout the development, showing a good understanding of their properties.

Q7 Recording decisions: They have put comments on the arrows throughout the development; these detail each decision and change.

Q8 Using graphic techniques: This candidate has sketched out some technical detail next to the models of their initial ideas. They have shown evidence of rendering, isometric, oblique and orthographic presentation within the development pages of the folio.

Q9 Using models: They have used quick sketch models to generate ideas. They have used accurate modelling within the development to evaluate the scale of the product and refine the aesthetics.

Q10 Planning manufacture: They have dimensioned sketches of each component and a cutting list that give sufficient detail about the product. The sequence also provides a structure for manufacture with details of equipment.

Q11 Time management: CANDIDATE 3 has invested considerable time in the development stage and would score highly in this area.

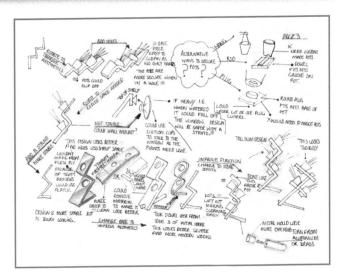

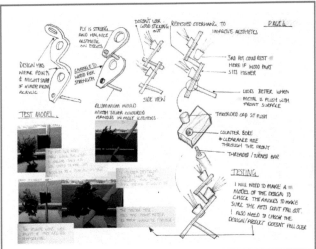

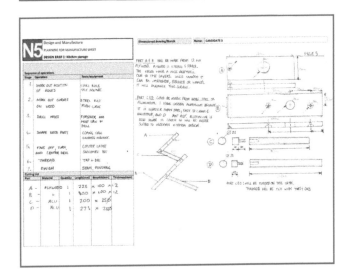

Candidate 4

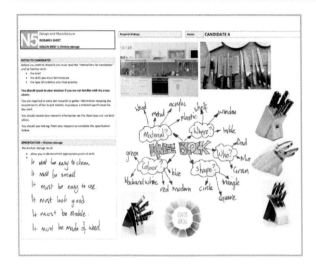

Q1 Analysing the brief: CANDIDATE 4 has created a mind map. This does not attract any marks as it isn't research. They have also collated lots of images of existing products and kitchens. This doesn't attract any marks as no information is gathered from it. Research further in the folio doesn't attract marks as it doesn't inform the specification. The specification doesn't score well as the points are not drawn from research.

Q2 Generating ideas: They have copied some existing designs, showing no creativity (see supporting evidence, which can be viewed via the weblink on page 16). All ideas are very similar and have little detail. Annotations are simplistic and do not add any additional information or clarity to the sketches.

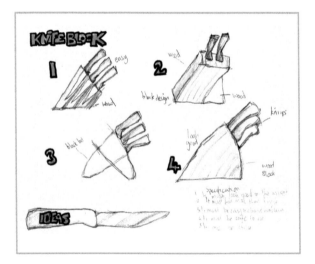

Q3 Exploring ideas: They have not demonstrated exploration of ideas. The candidate only collates information on materials and ergonomic data. Much of the information is not relevant and the candidate has not made any decisions. There are only a few token changes, mainly aesthetic, for example simple shape change. In addition, no reasons are recorded for any of the changes. The candidate develops a second idea that has resulted in repetition, showing similar but meaningless changes to the first development.

Q4 Refining ideas: They did very little meaningful exploration of their design, which has not really evolved as a result. There are limited details provided to permit manufacture. Annotations are basic, generic and show no real understanding of why decisions have been made or how well the product meets the specification.

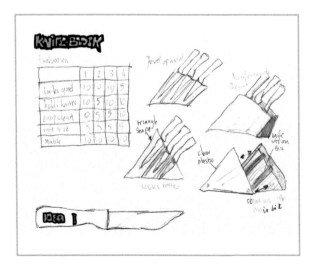

Q5 Using design factors: They fail to address any design issues in a meaningful way. They talk in simple terms about aesthetics and show no real understanding of ergonomics. Data is simply recorded.

Q6 Materials and manufacture: They have wasted time on some areas. Coloured headings, the knife logo, the evaluation table and repetition of rendering do not demonstrate appropriate skills, knowledge or understanding and would not attract any marks.

Q7 Recording decisions: The candidate has annotated each sketch like a new idea. There is no clear pathway, so the development is difficult to follow. Decisions are not recorded or explained.

Q8 Using graphic techniques: They demonstrate poor application of graphic techniques. The candidate has used a simple graphic style throughout the folio. The attempts at pictorial sketches do not demonstrate ability or understanding of any common methods.

Q9 Using models: This candidate has not used modelling techniques appropriately. They have created a model for the planning proforma that serves no purpose. The candidate's sketches communicate the same information.

Q10 Planning manufacture: The candidate has a basic dimensioned sketch that communicates some details about the solution. The cutting list only includes the wooden block. The steps in the sequence are vague and the tools are not related to the steps. The candidate would get some marks for the knowledge they have displayed.

Q11 Time management: CANDIDATE 4 has wasted time on some areas. Using evaluation tables not only takes up a lot of space, but is also time consuming. On their own, they will not attract any marks as they do not demonstrate understanding or reasons for decisions.

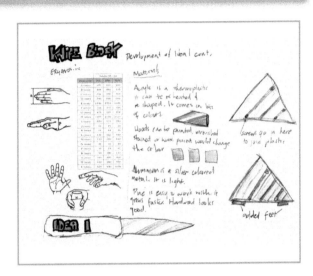

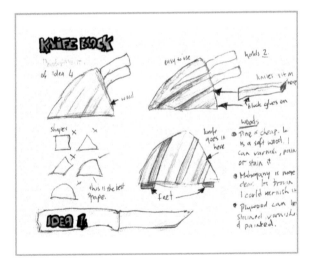

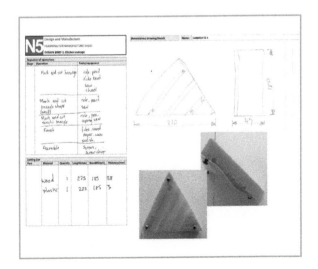

Manufacturing the Prototype

1. What can you manufacture for the final product?

You must accurately manufacture the product that you designed in Assignment Design. You should consider the time you have available and your skills in the workshop when designing the product.

You should ensure that your design solution will allow you to demonstrate and apply a high level of practical skills.

2. What will you do to ensure you achieve good marks for marking out?

You must ensure that you demonstrate your practical skills. Generating a template using CAD/CAM (computer-aided design/computer-aided manufacture) will not gain marks in this section as you will not be demonstrating your ability to use tools and equipment for marking out material. Templates, if appropriate, can be hand drawn to aid with marking out shapes. Having your material supplied cut to length could also limit your opportunity to mark out, depending on the complexity of your design. Consider what marking your project involves, such as: position of holes, shapes, joints and bends.

> **Assignment Tip**
>
> To merit full marks you must demonstrate your skills in accurately measuring and marking out parts with reasonable complexity.

3. How will the accuracy of your work be assessed?

Your teacher will use your working drawing to check the accuracy of your manufactured product. You will have to create a working drawing or dimensioned sketch to allow you to calculate the material you will need to manufacture your design.

The working drawing can be included in the planning for manufacture pro forma where it will attract marks for communication (graphic techniques). You must keep a photocopy of your plan for manufacture sheet to take into the workshop, as the original is sent to SQA with the rest of Assignment Design.

4. What will you do to ensure you achieve good marks for using hand and machine tools?

Any cutting should be done with accuracy, close to the line and without creating any damage to the surrounding faces of the material. Shaping should remove any traces of cutting, and be carried out with skill and accuracy to get the desired shape or fit required. Any forming should be planned and executed with care. Work should be formed with accuracy in line with the sizes in the working drawing. Using a range of tools and processes will give you an opportunity to demonstrate more skills.

> **Advice**
>
> Any components made using CAD/CAM will not attract any marks for cutting, as the machine has cut the parts for you.

Repetition of the same skill and tool use will only show ability limited to that activity. Remember this is a test and you want to show off what you can do.

5. What will you do to achieve good marks for the assembly of your product?

There must be an assembly. The marks awarded in this section vary based on the complexity and type of assembly. Simply screwing a hook into wood or butt jointing material would not allow you to demonstrate a high level of skill. The resources and methods you select for assembly should result in a functionally sound product. Remember, the Assignment is about demonstrating your skills. If you have cut two parts that now fit well together – the skill here is in the accurate cutting, not in the assembly – consider the number of stages and the equipment required to aid the assembly of your product. This can indicate complexity in assembly. Components should be set out as details in the working drawing or dimensioned sketch.

6. What will you do to ensure you achieve good marks for the quality of finish on your product?

You need to demonstrate a high quality finish. This means you must ensure you fully prepare your materials prior to assembly, removing any marks from marking out, cutting and shaping. There should be no evidence of glue, runs or blemishes on any applied finish, like paint, stain or varnish.

Finish, like all other aspects of your practical work, should be thought about in terms of complexity and quality. Again, any components finished using CAD/CAM would be *awarded no marks* for the finishing stage.

Assignment Tip

Finishing the edge of a simple plastic component would not be sufficient to merit top marks, as it would not demonstrate a high level of skill.

Evaluating

7. How will you evaluate your manufactured proposal?

You cannot carry out the evaluation until you have manufactured your solution. You will then need to test it to ensure it functions as intended.

The evaluation is not about your design process or how you manufactured the product. You must ensure that you evaluate the *manufactured product* against all of the specification points. A good evaluation will use research and evaluation techniques other than just personal opinion. You could also include any recommendations for improvements that you have identified.

CANDIDATES 1, 2 and **3** have all produced a detailed evaluation and would all score well. Candidate **4** has demonstrated no understanding.

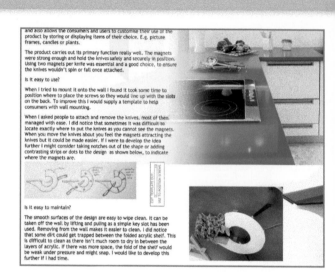

In this extract, Candidate 1 has evaluated all points from the specification in detail and also sketched some ideas for improving the design. They have tested the product and also asked people to try the product and provide feedback.

8. What supporting evidence can be submitted when your work is being sent to or verified by SQA?

Your teacher will submit your Assignment Design to SQA to be marked. You must ensure you have your double-sided research proforma, five A3 sheets and the planning for manufacture pro forma collated with your name on each sheet, ready to be sent away.

Make sure any pictures of models used within your folio are clear and that you have explained their use.

You need to remember to keep a photocopy of the planning for manufacture pro forma in school as you will need this to manufacture your product. Your teacher will also need to use this sheet when they are assessing Assignment Practical. If SQA come out to verify your practical work, they will also need access to your planning sheet.

Your teacher may talk to you prior to starting Assignment Practical. If this conversation results in any changes/improvements to your solution, these should be marked up only on the copy of your planning sheet that is to be kept at school.

If SQA come to look at your practical work, they will need to see the physical model, photographs will not be accepted. They will also be assessing your evaluation.

NOTE: Some activities carried out during the manufacture of your product will not attract marks. Ensure you read this section so you do not lose marks in the assessment of your practical skills.

- **The product:** If you manufacture something different from the product you have designed, it will *not* attract marks. The product you manufacture *must* be the one that you have designed.

- **Marking out:** Using a template created on CAD/CAM *will not attract marks*. This method of marking out does not allow you to demonstrate a high level of skill using the tools in the workshop. Using CAD/CAM to mark out material would also not allow you to demonstrate these skills.

- **Cutting and shaping:** Components produced using CAD/CAM *will not attract any marks* as these do not demonstrate your practical skills.

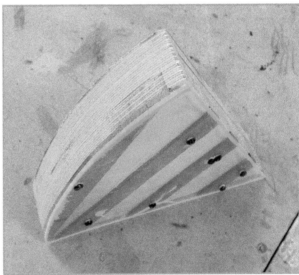

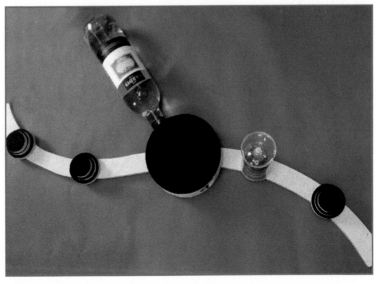

Understanding Command Words

IMPORTANT

When you sit the Question Paper, as well as having acquired the knowledge to answer the questions, you must also have the literacy skills to allow you to answer the different styles of questions correctly.

Before you attempt any of the questions in this book, it is important that you learn the skills covered in this chapter. Be aware that many questions require you to give an extended response, rather than a one-word answer.

In this section, you will become familiar with the different types of questions, and gain an understanding of the different command words and how to approach them. The command words you are likely to see in the Question Paper are explained below.

State

When asked to state, you must give a short factual answer. Similar command words, requiring a similar response, include name, identify, give or list.

Examples

Question: State an idea-generation technique that would be useful to a designer when designing a new chair.

Answer: *Morphological analysis.*

Question: Give a reason why this idea-generation method would be suitable to use when designing a chair.

Answer: *Morphological analysis is suitable because it is a structured method that would allow the designer to generate lots of ideas based on the different attributes of the chair. This technique will allow them to come up with ideas they might not have thought about otherwise.*

Describe

When asked to describe, imagine you are giving instructions or details to someone on the phone. Your aim is to help them visualise exactly what you can see. Don't just give a list. When describing, you must provide an account of characteristics, features and facts. You might consider questions involving 'who, what, where, when and how' to help detail your answer.

Question Paper Tip

Planning your answer by breaking it into manageable chunks can help when you are under pressure in the exam.

Example

Question: Describe how this idea-generation technique would be used by the designer when designing a chair.

Answer: *First, the designer would identify different attributes of the chair, using these as column headings in a table (e.g. material, number of legs, special features). They would then create lists of elements under these headings. The designer would select, at random, one element from each column. Next, they would use the random combination of words to inspire a new idea. This process can be repeated, giving a range of different combinations.*

Explain

When asked to explain, you should respond with facts and reasons. First, think about 'why' questions, then consider the events or details that impact the situation. Remember to make the relationships between things clear.

Example

Question: Explain why this idea-generation technique is useful for the designer during the chair-design task.

Answer: *Morphological analysis is useful because it allows the designer to list the different attributes of the product. By selecting combinations of attributes at random, the designer can generate many different creative and unusual designs.*

Answering Questions

This section will show you examples of both good and bad responses to exam-style questions from two candidates: Candidate A and Candidate B. Read the questions, then the answers from each candidate. Try to decide if they have understood the command words in the question and successfully written the answer. A summary of their success is given in the table shown opposite.

Example Questions and Answers

Q1: State the function of the sledge shown on the right.

Candidate A: *To transport the user downhill over snow.*

Candidate B: *The primary function is to transport the user downhill over snowy ground whilst sitting.*

Q2: The product is targeted at adult users. Give reasons for the designer's choices of aesthetics for this market.

Candidate A: *The seat and handles are black and the metal parts have been left showing the silver colour. The seat is rounded and the steering wheel is square.*

Candidate B: *The designer has chosen to use sophisticated colours (black and silver) that will be more appealing to an older market. Children would prefer bright coloured sledges. The product has a very functional look. The metal components and tread texture on the plastic parts make this sledge look expensive and well made.*

Q3: Describe how ergonomics has influenced the design of the sledge.

Candidate A: *Ergonomics has influenced the design as the seat has been shaped narrower at the front and is padded for comfort. The two foot rests are in front of the seat, allowing the user to sit in a comfortable position. The designer has sloped the footrests upwards. This improves ergonomics as it stops the user sliding forward when travelling downhill.*

Candidate B: *Ergonomics is all about how humans interact with products. Anthropometrics is to do with the sizes. Lots of sizes have been considered when designing this sledge. Physiology looks at the strength and effort required to use the product. Not much strength is needed to use this sledge.*

Q4: Modelling is an important part of the design process. Explain how modelling could have been useful to the designer during the development of the sledge.

Candidate A: *Modelling can be done in lots of materials and at different stages in the design process.*

Candidate B: *Modelling would have been useful to the designer in lots of ways. A scale model would have allowed them to work out the proportion and sizes of the sledge before making it full size. A CAD model would have allowed the designer to visualise the design in a range of materials that they could discuss with the client, without having to make a physical model. A part model could also have been used to allow the designer to test the steering components of the sledge and to examine how it might connect to other parts.*

How did they do?

	Candidate A	Candidate B
Q1	Both candidates, answers are correct in knowledge and are worthy of marks.	
Q2	This answer is not a justification. The candidate has simply described what parts look like but has failed to give any reasons.	This is a good answer where the candidate has made statements and correctly given reasons for their opinion.
Q3	This is a good example of a description. If you closed your eyes, you would be able to imagine the product. The ergonomic information in this answer is also detailed and clearly related to specific parts of the sledge.	This would gain no marks. The candidate only makes statements that define ergonomics. They speak generally about sizes and effort, but have not demonstrated that they understand how ergonomics has been applied specifically to the sledge. If you closed your eyes and listened to this answer, would you be able to visualise the sledge?
Q4	This would gain no marks. The candidate has only given facts about modelling. An explanation requires much more information or detail in the answer.	This is a perfect example of an explanation. The candidate clearly explains how three different modelling techniques would help the designer solve a variety of development problems.

Did you identify where the candidates had not responded correctly to the command words? Try reading over their answers again if you are unsure.

Area of Study 1: Design – Revision Checklist

Design Process

I can confidently draw on and apply knowledge and understanding of:

Got it?

- the **design team**, including designers, market researchers, accountants, engineers, manufacturers, marketing teams, ergonomists, economists, and their roles

- **identifying a design problem** through situation analysis, 'needs and wants' and product evaluation

- **design briefs** and the generation of **design specifications**

- **idea-generation techniques** such as morphological analysis, brainstorming, technology transfer, analogy and lateral thinking

- **development and refinement** of ideas through synthesis of ideas, justification and recording of decisions taken

- **product evaluation** by surveys, user trials, comparisons with specifications, the concept of function and fitness for purpose.

Design Factors

I can confidently draw on and apply knowledge and understanding of:

- **function** (primary and secondary functions, fitness for purpose)

- **performance** (ease of maintenance, strength and durability, ease of use)

- **market** (consumer demands, social expectations, branding, introduction of new products, market segments, marketing mix, needs, wants, technology push, market pull)

- **aesthetics** (shape, proportion, size, colour, contrast, harmony)

- **ergonomics** (establishing critical sizes, basic understanding of how humans interact with products, anthropometrics).

Communication Techniques

I can confidently draw on and apply knowledge and understanding of:

GOT IT?

- a range of **graphic techniques** (working drawings, annotated sketches and drawings, perspective sketches, illustration and presentation techniques, scale and proportion, simple orthographic drawings)

☐ ☐ ☐

- a range of **modelling techniques** (scale models, use of appropriate modelling materials or CAD models).

☐ ☐ ☐

Impact of Technologies

I can confidently draw on and apply knowledge and understanding of:

- the impact of design technologies on society and the environment (rise of consumerism, affordable and accessible products, and potential impact of design and manufacturing decisions on society and the environment).

☐ ☐ ☐

Area of Study 2: Materials and Manufacture – Revision Checklist

Planning for Manufacture

I can confidently draw on and apply knowledge and understanding of:

GOT IT?

- **preparing materials**, selecting appropriate **tooling and finishes** and reading **working drawings**.

☐ ☐ ☐

Tools, Materials and Processes

I can confidently draw on and apply knowledge and understanding of:

- a range of common **workshop tools** to measure, mark out, cut, shape and form materials

☐ ☐ ☐

- appropriate and acceptable **tools or equipment** to hold, clamp and restrain materials

☐ ☐ ☐

- a range of common and acceptable **machine tools** to sand, shape and drill materials

☐ ☐ ☐

- a range of **fixing and joining techniques** to joint/join woods, metals and plastics, including thermal joining and adhesive bonding

☐ ☐ ☐

- a range of **metal manufacturing processes** to cut, shear, notch, parallel and step turn, taper turn, drill, knurl, file, fold and bend metals. And, in industry, casting and die-casting

☐ ☐ ☐

- a range of **plastic manufacturing processes** to cut, drill, file, bend, twist and mould plastics. And, in industry, vacuum forming, injection moulding, and rotational moulding

☐ ☐ ☐

- a range of appropriate techniques and resources used to prepare and **finish** materials.

☐ ☐ ☐

Manufacturing in Industry

I can confidently draw on and apply knowledge and
understanding of:

GOT IT?

- **manufacturing in industry**, including CAM benefits, unit
  cost for mass production, quality assurance, globalisation,
  clean manufacturing

☐ ☐ ☐

- **rapid prototyping** technology

☐ ☐ ☐

- common **industrial processes**, such as injection moulding,
  rotational moulding, sand casting and die casting

☐ ☐ ☐

- **standard components**.

☐ ☐ ☐

Properties of Common Materials

I can confidently draw on and apply knowledge and
understanding of:

- types of **wood**, their properties and general uses (softwoods,
  hardwoods and manufactured boards)

☐ ☐ ☐

- types of **plastic**, their properties and general uses (thermo
  and thermoset plastics)

☐ ☐ ☐

- types of **metal**, their properties and general uses (ferrous and
  non-ferrous metals).

☐ ☐ ☐

Impact of Manufacturing Technologies and Activities on the World of Work and Society

I can confidently draw on and apply knowledge and
understanding of:

- reduction in workforce, skilled workforce, impact on
  environment (energy and pollution)

☐ ☐ ☐

- the measures that can be taken to support **sustainability**,
  such as the circular economy.

☐ ☐ ☐

Health and Safety

I can confidently draw on and apply knowledge and
understanding of:

- safe working practices and systems applicable to
  manufacturing activities, workshops or environments.

☐ ☐ ☐

Using the Revision Questions

The following chapters cover all the topics of the Design and Manufacture Course. The Question Paper will test you on a range of areas taken from these topics.

IMPORTANT

At the start of each chapter, you will find **revision questions**. These are designed to allow you to identify the areas that you particularly need to revise. These will test what you already know and highlight what you don't know.

At the start of the revision questions, we have given page numbers that will lead you to extra information on each topic. If you get stuck on a question, turn to the pages in the chapter to find the extra information that will help you answer the question.

You can answer questions from each topic in any order; it is completely up to you how you use this resource.

Remember that these are *not* exam-style questions. They are given to kick-start your revision by testing your current knowledge.

Do not forget to use the *GOT IT?* traffic-light system and Exam Checklist to record your progress as you go through the questions.

Design Factors Revision Questions

In this section, you will find questions that cover the Design Factors. Use these questions to test your current knowledge and understanding, to identify areas for further revision. You should be able to answer all of these questions by the time you are ready to sit the Question Paper. Answers are on page 134.

The following page numbers show where you can find additional information on this topic:

- **Function (F)** – Page 45
- **Performance (P)** – Page 46
- **Ergonomics (E)** – Page 48
- **Aesthetics (A)** – Page 52
- **Market (M)** – Page 54

1. **State** a **primary function** for the camping stove. (F)

2. **State** and **explain** a situation in which a camping stove could not be considered fit for purpose. (F)

3. In the table below, **explain** how the following aspects influence the performance of a product. (P)

Aspect	Influence on a Product
Value for money	
Planned obsolescence	
Ease of maintenance	
Durability	

4. **State three** ways in which the designer can ensure a product is easy to maintain. (P)

5. **List** the advantages and disadvantages of planned obsolescence for (P):

   a) the consumer

   b) the manufacturer.

6. **Describe** how shape and form has influenced the design of the hair dryer shown on the right. (A)

7. Not including shape and form, **list six** elements that influence the aesthetics of a product. (A)

8. In the table below, **explain** why each ergonomic aspect would have been consideration by the designer when developing the products listed. (E)

Aspect	Product	Explanation
Reach	Shelf	
Clearance	Desk	
Grip	Bike	
Weight	Elevator	
Strength/effort	Medicine Bottle	
Body dimensions	XBOX controller	

9. **Give two** examples of how anthropometrics was used to influence the design of the garden hand tool shown on the right. (E)

10. **Give** an example of how a designer could ensure a crash helmet would fit a wide range of users comfortably. (E)

11. Market is influenced by various elements. **Explain** what is meant by the following terms: (M)

    a) consumer demand

    b) market segment

    c) technology push

    d) needs and wants

    e) branding.

Function

Primary and Secondary Function

All products have an intended purpose. The **primary function** is what the product is designed to do. Many products also have **secondary functions**. These are spontaneous alternative uses, like using a knife to open a box, or planned secondary functions that can make a product more appealing.

In the Question Paper, you might be shown a product and asked about its function. Function is also important for the Assignment, as you will have to have an understanding of the concept in order to successfully develop the product.

A product could be considered '**not fit for purpose**', depending on the situation.

Fitness for Purpose – Case Study

The three products shown have the same primary function – to boil and dispense water. However, each could be not fit for purpose, depending on the situation.

Canteen

A kettle boils water and is designed specifically for domestic use. A kettle would not be fit for purpose in a work canteen as it would not hold enough water. In this instance, an urn would be used. It has a tap, so does not need to be lifted for pouring. It holds lots of water and can regulate its temperature over time. An urn could be described as 'fit for purpose' when used to give a continuous supply of hot water in a canteen.

Camping

The domestic kettle would be impractical when camping as it is bulky and requires access to electricity. The urn would be similarly impractical. The camping kettle is fit for purpose because it is light, compact and can be used on an open fire or camping-gas cooker.

> **Advice**
>
> When designing a product, function is a good place to start your development.
> You will need to resolve how the product works and ensure it is fit for purpose. Exploring the function will naturally affect other design factors, such as ergonomics and aesthetics.

Quick Test

1. What is a secondary function?

2. What is meant by 'fit for purpose'?

Performance

When shopping, consumers compare products to ensure they get value for money. Consumers are looking for products that are:

- easy to use
- easy to maintain
- durable.

Ease of Use

Products should be designed to make life easier. If a product is difficult to use, consumers will not be satisfied. Clear instructions, ease of assembly and good ergonomics all impact ease of use.

Ease of Maintenance

A product is considered easy to maintain if it is easy to clean and repair, and to fit replacement parts. A bicycle is an example of a product that is easy to maintain. During normal everyday use, parts such as tyres and chains may become worn or damaged. The use of replacement standard components increases the lifespan of the product, ensures it is easy to maintain and provides consumers with **value for money**.

Not all products are easy to maintain. Xbox controllers are a good example of a product that is difficult to maintain. The manufacturer does not use standard screws to assemble the product. Screws that require a special tool are used.

Durability

Not every product is designed with durability in mind. For example, if a consumer bought a pair of earphones for £1, they would not expect them to last as long as a set that cost £30.

The following aspects can affect the durability of a product:

- material selection and construction
- use and abuse of the product.

Durability is affected by how the user uses (or abuses) the product. Swinging on a chair may weaken or break the back legs, shortening its life span. Dropping a product in water or forgetting to switch appliances off can also affect their performance.

Durability determines how long a consumer keeps a product. This can have either a positive or negative effect on the environment.

A high turnaround of products can be good for the economy and for the manufacturer (if it ensures repeat business). However, this has a negative impact on the environment as the unwanted or broken products are sent to landfill.

Planned Obsolescence

Planned obsolescence is a key factor in performance. Designers do not design products to break; however they may design a product to have a limited lifespan. This is called planned obsolescence and is used because:

- products go out of fashion
- new technologies emerge
- components wear out and can't be replaced.

When products are designed to have a limited lifespan, cheaper materials and components can be used.

Obsolescence means that consumers sometimes throw away products that are still in good working condition, replacing them for better or more advanced versions.

There are advantages and disadvantages of planned obsolescence to both consumer and manufacturer.

	Consumer	Manufacturer
Advantages	**Cheaper to buy** – manufacturers can make the products for lower cost. **Continual change** provides more choice.	**Increased sales** as the market doesn't saturate. **Updates** may increase brand loyalty and awareness.
Disadvantages	Consumers can be exposed to **peer-pressure** and may feel compelled to buy a newer, better product, even if the one they already have is in good working condition. **Costs** – consumers need to spend more money regularly as the frequency of updgraded products on the market is high.	Continual research and development costs. Repeated risks associated with product launches.

Quick Test

1. **List** the things that affect the performance of a product.

2. How can planned obsolescence affect the environment?

Ergonomics

Ergonomics is the study of how humans interact with the objects and spaces around them. Designers must apply their understanding of ergonomics to create products that are comfortable, safe and easy to use.

In the Question Paper, you may be asked to identify how ergonomics has influenced the design of a product. During your design work for the Design Area of Study and Assignment, you may choose to use **data tables** and/or **models** to solve ergonomic issues within your idea development.

You should consider these fundamentals when exploring ergonomics:

- body dimensions
- physical ability
- limitations of movement
- user interactions.

Question Paper Tip

Always be specific when describing/explaining ergonomics.
For example, stating that a chair should be the correct size is too generic. Instead, you could state *the width of the chair should accommodate the 95th percentile user.*

Anthropometrics is the study of human sizes. Dimensions of men, women and children of all ages have been gathered over years of research. This information can be found in anthropometric data tables. When developing a design, the data selected would depend on the **target market** and the specific sizes relevant to the product. Sizes can be tested and evaluated through modelling.

Using Anthropometric Data

There is no such thing as an 'average person'. Let's consider an example.

For a doorway to be usable by the majority of the population, the designer would have to identify **critical sizes**. In this case, the height and width of the doorway are important.

As shown in the data table below, the 95th percentile male represents the tallest of the target market, the 95th percentile female represents the largest hip width and the 95th percentile male represents the largest shoulder width. In the design of the doorway, the designer would use these maximum dimensions to establish what size the doorway should be; allowances would be made to ensure clearance for outside clothing and footwear. The typical dimensions of wheelchairs and large items of furniture would also influence the size of a doorway.

Anthropometric Data Table

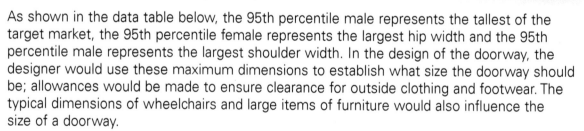

	Men (Percentiles)			Women (Percentiles)		
	5th	50th	95th	5th	50th	95th
Height	1630	1745	1860	1510	1620	1730
Shoulder Breadth	370	405	440	340	365	390
Hip Breadth	350	390	430	350	405	460

Other than the sizes, people differ in their **physical abilities** and **limitations of movement**.

Some people have limited physical abilities that can affect their capacity to use products. The designer has to take into consideration the **agility** and **strength** of different users to ensure the product is accessible to as wide a market as possible.

When a designer is developing a product, they need to consider how users will **interact** with it by identifying any aspects that would make it difficult, dangerous, confusing and/or uncomfortable to use.

Task 1

In the cordless drill shown below, four features have been highlighted for which ergonomics were considered. Use the ergonomics table (right) to identify and colour code these features in the lower table. An example has been given for you.

(Answers on page 140).

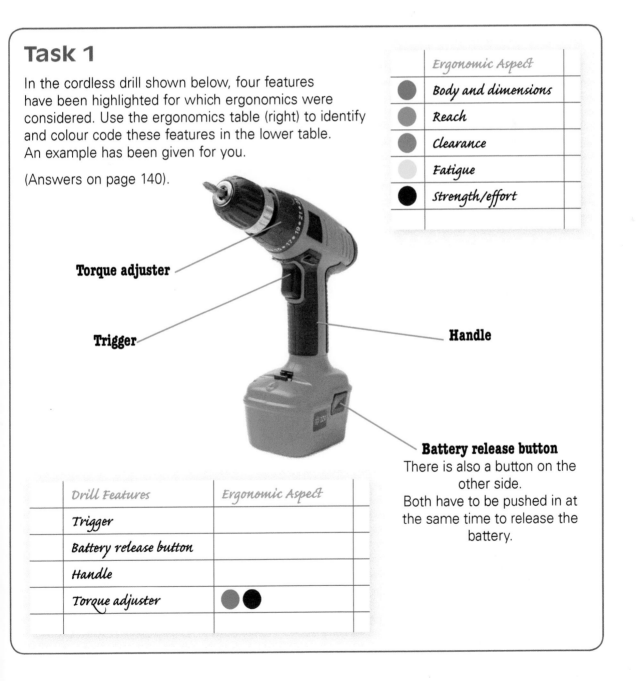

Ergonomic Aspect	
●	Body and dimensions
●	Reach
●	Clearance
○	Fatigue
●	Strength/effort

Torque adjuster

Trigger

Handle

Battery release button
There is also a button on the other side.
Both have to be pushed in at the same time to release the battery.

Drill Features	Ergonomic Aspect	
Trigger		
Battery release button		
Handle		
Torque adjuster	● ●	

Task 2

	Men (Percentiles)			Women (Percentiles)		
	5th	50th	95th	5th	50th	95th
Hand Length	179	193	211	165	180	205
Hand Breadth	84	90	98	73	79	86

1. What sizes from the data table would be used to inform the size of the handle of the iron?

2. **Explain** why the designer would have considered the weight of the iron when refining the ergonomics.

3. When designing the iron, the designer made a model. Why is modelling a good method to test reach? (Answers on page 140.)

Task 3

Trigger buttons

Dual-shock joysticks

Explain how ergonomics has influenced the design of the controller shown above (answers on page 140). In your answer, you should consider the:

- form of the controller handles
- position of the coloured buttons
- spacing between buttons
- position of the trigger buttons
- size of the dual-shock joysticks.

Task 4

Some different features of a bicycle have been labelled.

Use the ergonomics table (right) to identify and colour code these features in the lower table. An example has been given for you. (Answers on page 140.)

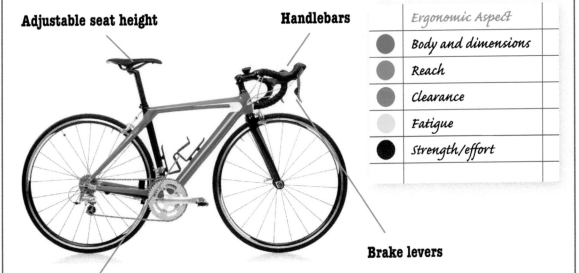

Adjustable seat height

Handlebars

Brake levers

Water-bottle holder

	Ergonomic Aspect
●	Body and dimensions
●	Reach
●	Clearance
●	Fatigue
●	Strength/effort

	Drill Features	Ergonomic Aspect
	Adjustable seat height	
	Water-bottle holder	
	Handlebars	
	Brake levers	● ● ● ●

Aesthetics

In the Question Paper, you may be asked how **aesthetics** has influenced the design of a product. You will also require an understanding of aesthetics to allow you to develop and refine your ideas within both the Areas of Study and the Assignment.

The term 'aesthetics' relates to our response to objects that we consider to be beautiful or pleasing. A consumer's first impression can influence their choice of whether or not to buy a product.

It is important for the designer to understand the preferences of the target market. What one person finds attractive, another may not. We all have our own opinions about style.

The aesthetics of a product are influenced by a number of elements, including:

- shape and form
- proportion
- colour
- texture
- materials
- contrast and harmony.

When describing aesthetics, it is important to mention the aesthetic elements listed above and to evaluate the effects they have on the product.

Example Question

Watch 1

Watch 2

Compare how colour and proportion have been used in these watches.

Answer:

Watch 1 uses sophisticated colours. The black and silver colours contrast, making the dials and time easy to see. The watch face is larger than the strap. The different dials are different sizes, relating to their importance.

Watch 2 is fun-looking because of the blue colour. The silver-coloured metal adds value to the watch. The strap and watch face are the same width, in keeping with the simple style.

Task 1

Compare how colour and pattern have been used in the suitcases shown here.
Answers are given on page 140.

Task 2

Compare how shape and contrast affect the design of guitars shown below.
Answers are given on page 140.

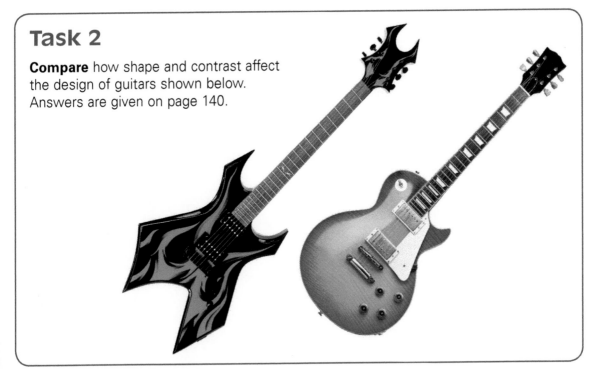

Quick Test

1. **State five** aspects that can affect the aesthetics of a product.

2. Why is it important for a designer to consider aesthetics?

Market

The market can be defined as a group of people who are potential users for a product or service.

Market Segments

There are numerous ways to split a market into **segments** (or sections), for example:

- **Geographic factors**, such as country of origin or residence, language and culture, are important in matching the product to the market.
- **Economic and demographic factors** also affect people's needs and interests. These include differences in age, income, gender, education and social status.
- Consumers may also buy into a product due to **personal motivations** such as peer status, attitudes, lifestyle, goals, media and expectations.

There will still be differences between consumers within a market segment.

Target Market

Within a market segment, for example single adults aged between 30–45, there will be males and females with a variety of lifestyle needs and wants. It would be difficult for a designer to design a product that would suitable for everyone within this broad market segment.

A target market is usually a more specific group of people, for example, brides-to-be, teachers or gardeners.

Niche Market

A niche market is a narrowly defined part of a target market. For example, consider a target market of cyclists. There are hundreds of products aimed at this target market, such as frames, seats and bells. An example of a market niche within this target market is commuting cyclists. Pannier bags and folding bikes are aimed at this niche market.

Task – Branding

Branding plays an important part in consumer expectations and influences how likely we are to buy a product.

With reference to well known luxury and budget brands, **discuss** the following points. (You should carry out some research to help you answer this task.)

- Consumers can be **loyal** to a successful or reliable brand.
- Branding can affect consumers' **perception of the quality** of a product.
- Branding can satisfy consumers' social and emotional needs.

Quick Test

1. What is the difference between a niche market and a target market?

2. How does branding influence consumer choice?

Design Process Revision Questions

In this section you will find questions that cover the design process. Use these questions to test your current knowledge and understanding, and to identify areas for further revision. You should be able to answer all of these questions by the time you are ready to sit the Question Paper. Answers are on page 134.

The following page numbers show where you can find additional information on this topic:

- **Design Team (DT)** – Page 58
- **Design Process (DP)** – Page 60
- **Design Brief (IP)** – Page 61
- **Research (R)** – Page 62
- **Specification (S)** – Page 66
- **Communication (C)** – Page 68

- **Modelling (M)** – Page 70
- **Idea-Generation Techniques (IG)** – Page 75
- **Development and Refinement** – Page 80
- **Evaluation (E)** – Page 84
- **Proposal (P)** – Page 86

1. **Complete** the table below, explaining the role of each member of the design team and saying how they support the designer. (DT)

Member	Role	Support
Designer	To create design solutions	
Market researcher		
Accountant		
Engineer		
Manufacturer		
Ergonomist		

2. **State** the name of **two** techniques that a market researcher could use to find information on a target market. (IP)

3. **State four** methods that the marketing team could use to promote a new product. (DT)

4. What methods could the designer use to communicate with the manufacturer? (C)

5. Why is analysing the brief an important stage of the design process for the client and for the designer? (P)

6. **Explain** why it is essential, after analysing the brief, to carry out initial research. (R)

7. In the table below, **give** an example of an aspect of a product that could be researched, making appropriate use of the different techniques. (R)

Research Technique	Research activity/purpose
Testing	How long it takes a kettle to boil water
Questionnaires	
Search engines (online)	
Measuring and recording	
Using data	
User trip/trial	

8. **Explain** how poor or no research can impact the specification. (R)

9. **Explain** the purpose of a product design specification. (S)

10. **Describe two** ways in which the designer would use the specification at different stages of the design process. (S/E)

11. **Describe** how the following methods can be used by a designer to generate and develop ideas (IG):

    a) technology transfer

    b) biomimicry

    c) SCAMPER

    d) six-hat thinking.

12. **State** an advantage and a disadvantage of using the idea-generation method '**Taking your pencil for a walk**'. (IG)

13. **State** an advantage and a disadvantage of using **existing products** as inspiration for new design ideas. (IG)

14. Briefly **explain** how evaluation could be useful throughout the design process. (E)

15. **State three** aspects of a design that could be **evaluated** or **communicated** through modelling. (M)

Design Team

In industry, production of a successful product requires the expertise and input of a range of people, including the design team. You should familiarise yourself with the roles of these experts and the relationships between the various team members. You must understand when and how they contribute to the successful development of a new commercial product.

Designer

It is the designer's job to create solutions by working out how products should function and styling them to best suit the needs of the client. The designer is responsible for generating the idea, as well as producing detailed drawings or models that can be used to communicate ideas to the rest of the design team.

Market Researcher

Market researchers are important as they can communicate with the consumers on behalf of the designer. Market researchers provide information that helps the designer understand the market they are designing for.

The researcher may need to investigate these market features:

- gender and age range
- geographic location (country/area/global)
- disposable income
- lifestyle (hobbies and interests).

The market researcher might also identify a gap in the market for a specific product or service – the **market niche**.

Accountant

During a design project, money may be required for research and development, testing, manufacturing and advertising. The accountant's role is to look after finances. They advise the design team on budgets, as well as monitoring money spent and money earned.

Engineer

The engineer works with the designer to ensure that the new product can, in fact, be manufactured successfully. The engineer advises the designer, ensuring materials, components and structures are suitable for safe and functional use of the new product.

Manufacturer

The designer will work with the manufacturer to select the best way to produce the product, depending on the cost and volume of production. Manufacturers use working drawings to reproduce accurately the designer's idea. They communicate with designers and engineers to ensure that the materials, assembly and form mean that the product is safe and fit for purpose.

Marketing Team

The marketing team is responsible for bringing the product to consumers' attention and enticing them to buy it. There are many strategies the marketing team can use to do this, including:

- TV, radio and magazine advertisements
- using celebrities to endorse a product
- promotions, giveaways and other incentives
- packaging and point of display (in store).

Ergonomist

The ergonomist looks at how we, as humans, interact with products and the world around us. The ergonomist advises the designer on aspects relating to our use of products and typical human sizes. It is the ergonomist's specialism to identify correct clearance, reach, strength and size, and also psychological aspects to ensure the product is safe, comfortable and easy to use for the intended target market.

Economist

The economist keeps track of what is happening in the economy and determines how this might affect the consumer's desire for or ability to buy a new product. For example, if the cost of living rises, consumers on a lower income might struggle to buy food. If a family cannot afford to eat, they certainly will not pay a lot of money for new luxury products. The economist will also predict trends and patterns in the marketplace.

Consumers and Retailers Also Inform the Work of the Design Team

Although consumers are not directly involved in designing or manufacturing a product, the designer must consult with potential customers and understand the consumers' (target market's) needs. If the product does not meet the consumers' expectations, needs, financial constraints or appeal to them aesthetically, it is unlikely they will buy the product.

Of course, the retailer sells the product. They can also inform the designer of current trends, historic and predicted volumes of sales and market demand for new products. The retailer will work with the marketing team to agree on suitable point of display, which can help promote sales of the new product.

Identifying a Problem

The first stage of the **design process** is to determine a design opportunity. A design opportunity may be identified in a variety of ways.

You should familiarise yourself with the methods given below.

Question Paper Tip

In the Question Paper you could be asked to **describe** how a problem or opportunity might be identified.

Situation Analysis

Situation analysis is a research activity in which a situation or environment is studied. For example, people may be observed interacting with a specific product to identify any difficulties the users may have. This could reveal an opportunity for a better or improved product.

Needs and Wants

'Needs' and 'wants' are important features of the target market. **Needs** are essential requirements like food, clothing and shelter. **Wants**, on the other hand, are luxuries. Consumers can live without luxuries and they will not spend money on luxury products if they cannot meet their basic survival needs. Regardless of the intended market, the designer must know what consumers need and want. A designer might find this out through market research, speaking to clients or approaching retailers about sales and demand. Consumers have more choice now and the designer needs to meet their demands to sell products.

Product Evaluation

Product evaluation occurs when designers try to identify problems or opportunities by **evaluating** existing products. They do this by comparing and testing products, observing other people using them and recording feedback about any relevant design issues.

Quick Test

1. **State three** ways to identify a design opportunity.

2. **State two** ways to evaluate a product.

Design Brief

The **design brief** is the starting point in the design process. A good design brief should:

- provide the designer with a clear statement of the problem or the situation that demands a new product

- explain what is required from a solution

- clearly indicate the identity of the target market.

The brief is often written by the **client**. The client may describe a situation to contextualise the design problem.

The designer must explore the design brief to ensure they fully understand the design task.

You will be given at least one design brief for the Area of Study and the Course Assignment. In order to fully understand the design task you have been given, you will need to **analyse the brief**. There are several ways you could do this.

Creating a 'mind map' will allow you to explore and record different aspects of the brief. You could analyse the brief in the box above by asking yourself a series of questions, such as:

- **What is it going to be used for?** (Where will it be used/kept? Will it need to adjust? What do the users want?)

- **What is the Creative Currie range?** (What are the aesthetic characteristics of the range – colour, shape, form, style?)

- **What size is a 28 watt bulb?** (Are there different types of these bulbs? Are they different sizes?)

Brief

Infinity Design wants to add a new lamp to their Creative Currie product range. The lamp must be suitable for a teenage market and retail for no more than £20. Each lamp will be supplied with a 28 watt bulb.

Advice

Note, a mind map is not research and should not be used on the research pro forma for your Assignment.

Quick Test

1. What is the purpose of a design brief?

2. Why is analysing the brief important?

3. **Explain** how you could analyse a brief.

Research

The majority of commercial products that fail do so because of a lack of appropriate or meaningful research. If the designer does not understand the consumer and their needs, a product may not appeal aesthetically to the target market, could be difficult to use or otherwise unfit for purpose.

What Do I Research and Why Should I Do It?

Research is an essential part of the design process. It is a learning activity that will allow you to collect facts and information to incorporate into your design development.

There is no magic formula for what you should research. Every design problem is unique. What you research will depend on:

- your design problem
- any restrictions you need to consider
- what you already know
- what needs you have identified.

If you are given a design brief, you must research to clarify aspects of the brief so that you can write a *meaningful* **product design specification**. The specification requires *detailed* facts from your research.

Further research will become necessary as you progress through the development stage. For example, you may need to find out what materials are suitable to enable specific components to be manufactured or you may need to find out about specific fixtures, fittings or other standard components to be used in the assembly.

ADVICE

You will be required to use a range of research techniques to pass the Assessment Standard within the Design Area of Study.
Ensure you only research and display <u>useful</u> information. For example, you do not need to state the atomic number of materials!

Question Paper Tip

You may be shown a commercial product and questioned about what design factors the designer should research. You could also be asked to describe a research activity.

Information gained from research should be used to explore and refine the details of the design into a functional solution suitable for manufacture.

Natural Born Researcher

Research and investigation is something you have been doing every day since you were born and probably without realising it! You are already familiar with many research methods and, every day, are subconsciously selecting the most appropriate ones.

Consider the following: if you want to know what tops are trendy, you would look in stores; if you want to know if shoes are comfortable, you could try them on; if you want to find out how to spell a word, you would look it up or ask someone.

When you are researching, it is crucial that you use an appropriate source to give you the most reliable information. It is also vital to filter the information by extracting only the *relevant* bits.

Task – Research

(Answers on page 141)

Let's consider the teenage lighting design brief from page 61. The analysis on page 61 generated a list of things we would like to know more about. Here are some things you might consider researching before writing a specification:

- how teenagers might use the light
- the size of a 28 watt bulb
- different types of light switches
- how much teenagers might pay for the lamp
- the style of the Creative Currie existing range
- the aesthetic factors that appeal to the teenage market
- the different woods, metals and plastics that are available.

Before you spend time on research, you need to ensure that there is a reason for doing it. Ask yourself:

- How will I use my research results during development?
- How will the research inform my design decisions?

> **Research methods:**
>
> Questionnaire/survey
> Internet
> Books/magazines/
> journals
> Observations
> Testing
> Data tables
> Measuring
> Asking experts

Think about the relative importance of the potential research areas above and how the results would influence the design of the light.

1. Which **three** points would not lead to **meaningful research** *at this stage*?
2. What **design factors** do the above points relate to and why are they important?
3. Which **method**, from the boxed list on the right, would provide the most meaningful results for each of the research areas identified?
4. Why have you selected these methods over the others?

Quick Test

1. **List six** different research methods.
2. What is the purpose of research?
3. How can research influence the success of a product?

Examples of Research

There is no need to carry out masses of research, as long as you find out the information you require. Bear in mind that it is very easy to make assumptions, especially if you feel you know something about the **brief** or the **market**.

Let's compare two candidates' research for the teenage lighting task. How well do you think they have done?

> **Link to research work by Candidates A and B**
> https://collins.co.uk/pages/scottish-curriculum-free-resources

Candidate A

Candidate B

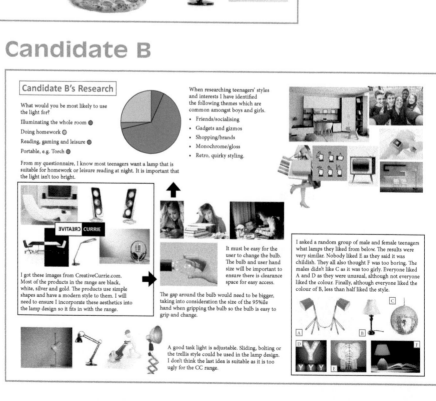

At a glance, both of the pages look very good. Remember research should be about finding out *relevant* facts. These facts help the designer to write up a detailed **design specification** for the next part of the process.

Candidate A appears to have made assumptions that the target market like football and games. This board is more about opinion than fact. It is likely that the candidate felt that they knew about this market and is thinking about designing for themself rather than bothering to research the **target market**. They have addressed most points raised from the analysis. However, they have been unable to draw conclusions at this stage. The manufactured board research is not relevant. Why would you use hardboard for a lamp? There is no value in writing about materials during your research, unless the information will be used when developing the product.

Candidate B has cleverly covered all points from the analysis. **Clear conclusions** are evident, which can later be recorded in the specification. They have used a range of valid research techniques. The results will be useful to them during development and will inform their design decisions.

	Candidate A	Candidate B
What will the teenager use the light for?	Not clear. Ceiling light and lamp shown.	Survey/questionnaire has been carried out. Clear results and decisions made.
What size is a 28 watt bulb?	Very good evidence. Bulb sizes are useful for development later on.	Clearance space around the bulb considered but no bulb sizes.
What aesthetic factors will appeal to the teenage market?	Again unclear. Could perhaps pick colours from images but that is all.	A survey of different lamps carried out: unusual and interesting lighting, monochrome, gloss, gadget and retro.
What is the style of the Creative Currie range?	No research on Creative Currie range.	Clear research and conclusions made about the style/aesthetics of brand.

Quick Test

1. **Explain** why you must be careful when researching materials.

2. Why is it important to draw conclusions from your research?

3. Why is the designer's opinion not sufficient research on its own?

The Specification

A **specification** is a list of all the features and functions of a product. A good **design specification** states all of the criteria that the final solution must meet, to satisfy the client's requirements. All of the factors that influence design are normally considered. Throughout the design process, the designer will refer to the specification, **evaluating** their decisions and ideas to ensure they are on the right track to a good solution.

Writing a Specification

Your specification must be based on your research. Generic statements that lack detail or clarity should be avoided. If you have carried out *meaningful* research, you will have identified several facts that you would not have known from simply reading the **design brief**.

Simplistic statements, such as '*it must be durable*', are too vague as they could be applied to any product. Look at the following extracts from a candidate's specification for the lighting task (page 61). Under each of the candidate's statements, the text explains why these points are unsatisfactory.

- *The product must produce light.*

The research should have clarified what type of light the target market wanted. A torch provides light. Would that meet the needs of the user?

- *The product must be attractive to teenagers.*

The research should have identified what styles teenagers find attractive. Everyone has different tastes. Colours, shapes, styles or themes could have been given.

- *The product must be easy to maintain.*

The research should have explored what features would make the product easy to clean or maintain. For example, it should use standard sized bulbs and have easy access to replace the bulb.

- *The product must hold a standard bulb.*

There are hundreds of bulbs on the market. The research should have explored the possible sizes of a 28 watt bulb. More specific information could have been given.

- *The product must be safe.*

The research should have explored issues relating to safety of lighting. For example, electrical safety and heat resistance of materials.

Although the candidate's statements in red are correct, they fail to offer the designer any real guidance. Each of these statements could easily have been written after reading the brief and without carrying out any research. The statements are vague and no meaningful or specific points have been made.

In the following extracts, the candidate has correctly used their research to inform and develop their initial attempt at the specification. Read over the improved specification points and the explanations of why they are now successful.

- *The lamp must be adjustable, so it is suitable for use when reading and writing.*

The candidate has been able to specify more detail about the function of the lamp.

- *The product must be silver and black to suit the Creative Currie range.*

The candidate has researched the range and identified important aesthetic attributes.

- *There must be a minimum of 20 mm clearance space around the widest part of the bulb to make it easy to access and replace.*

The candidate has given specific statements based on ergonomic research to clarify what would make the product easy to maintain.

- *The product must hold a 28 watt bulb with a standard E14 screw fitting.*

The candidate's research has allowed them to identify the most common and easy to use fitting.

Specification

1. Function

1.1 The angle of the lamp must be adjustable to suit reading and writing tasks.

1.2 The lamp must use a 28 watt bulb with an E14 fitting.

1.3 The lamp must include a cable tidy to store excess cable.

2. Aesthetics

2.1 The lamp must be black and silver to compliment the Creative Currie range.

2.2 The lamp must have a gadget/technology appeal to the teenage market.

2.3 The fittings must be hidden to ensure they don't spoil the look of the product.

Assignment Tip

In the Assignment, SQA provide a number of starting points in the briefs. You will get one mark for taking these into the specification points. Additional marks for the specification come from summarising the findings of your research.

Quick Test

1. Why is the specification an important tool for the designer?

2. How can you write a good specification?

Communication

Good communication and decision-making skills are essential for a designer. You will have to demonstrate communication skills to pass the Area of Study and the Course Assignment.

A designer uses communication skills:

- to discuss **initial ideas** with the client at an early stage.
- during **development** to present refined concepts to the client and to discuss the manufacturing options with the manufacturer.
- when producing the **Proposal**. The Proposal outlines the final design solution to the client, sets out the target market for promotion, and communicates assembly and technical detail to the manufacturer for production.

During the Design and Manufacture Course you need to demonstrate your ability to communicate your ideas and decisions to your assessor.

You should demonstrate that you have a range of communication skills and knowledge of when, why and how to apply these skills appropriately and effectively.

Drawing is only one form of communication and it is not enough on its own. The following pages in the communication section of this Success Guide will explain the range of techniques you could use to enhance or improve your communication skills.

When communicating, it helps to think about these questions:

- Why is the communication needed?
- Who is the communication aimed at?
- What information needs to be communicated?
- What method/type of communication would be most effective?
- What additional information needs to be communicated?
- What changes or decisions need to be communicated?
- What positive or negative points need to be communicated?
- What reasons for decisions and changes need to be communicated?

> **ADVICE**
>
> Use these different communication techniques at appropriate stages in your design work:
> - annotations and evaluations
> - sketching, 2D and pictorial
> - detailed views
> - renderings
> - modelling (take photographs).

Annotations and Evaluations

Sketches can only communicate so much information on their own. Good **annotations** are essential and add clarity to sketches. They can make it easier to understand any design work, regardless of the quality of sketching.

The reasons for decisions should be clearly based on research and should relate back to the specification. Written summaries or evaluations can help when recording the decisions that have been made.

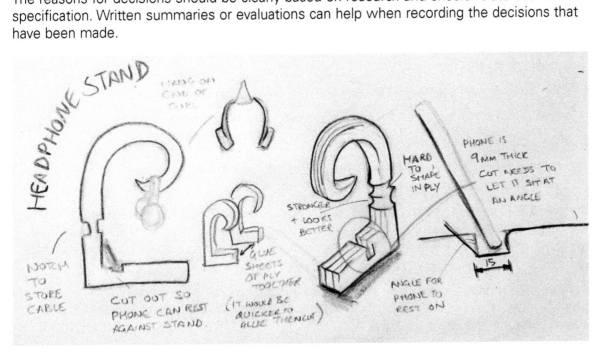

Sketching: Pictorials and 2D

If you are not great at sketching, you can use squared, isometric or perspective paper to help you. Sketches should show material thickness, be in proportion and pictorials must be in a known drawing style. 2D sketches are useful to show different views and are required in the planning for manufacture.

Detailed Views

Producing detailed views can involve scaling up of small parts of the design or showing the product from different positions (for example, a box opened and closed). Showing a detailed **exploded view** of some components can help to communicate how the product will be assembled. Exploded views can be 2D or pictorial.

Rendering

Any rendering that you apply should enhance your design work. If done well, it should communicate form, texture and materials. Rendering can be applied manually or digitally using 3D modelling or illustration software. It is possible to achieve full marks with no rendering.

Modelling

Modelling is brilliant when you can't quite sketch what you want to show.

Modelling should be used to communicate, problem solve, test or generate ideas. You must ensure that models in your folio have a purpose.

This means they should:

- communicate something that your sketches don't already communicate

- allow you to understand/learn something new about your design

- allow you to explore, visualise new ideas.

Remember you will also need to **annotate** the photographs of your models, so your thinking is clearly communicated.

Physical and CAD models can be used to communicate **scale**. You can do this easily by using silhouettes to give an idea of scale (as in the two-man tent presentation below) or by fitting the product into a suitable background image using some basic DTP, or by including something in the photograph with the model that identifies scale.

You can trace around and work into a photo of a sketch model if your graphic skill is not great. Twisting, forming and moulding and joining simple modelling materials is also a nice way to generate and explore ideas.

In the above example, the candidate has made a quick cardboard model of a tent. They have taken photographs of the model from different views and sketched on top of them. The use of silhouettes shows the scale of the model in relation to a person.

The orthographic layout in the corner helps to show what the design looks like from different view points.

Assignment Tip

Using modelling to generate initial ideas will allow you to attract marks for ideas in the modelling section in the Assignment. Throughout the remainder of your Assignment, you should use modelling to test something, work out a size or when you are having difficulty sketching it.

Task – Communication

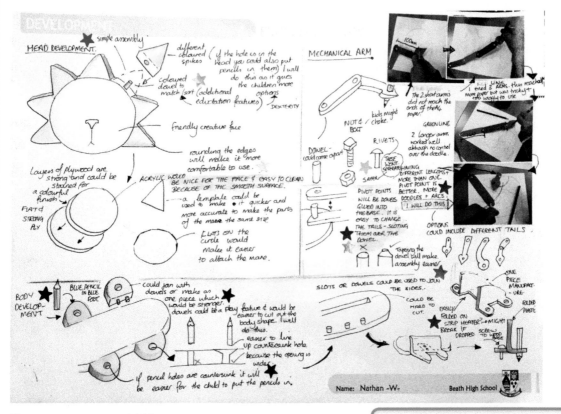

Name: Nathan -W-   Beath High School

(Answers on page 141.)

An extract of a candidate's design work for a toy is shown above.

Link to design work for a toy
https://collins.co.uk/pages/scottish-curriculum-free-resources

1. What communication methods has the candidate used?

2. Are the graphics that are used appropriate?

3. Have they recorded their decisions clearly?

4. Is there clear evidence that the candidate has justified any decisions they made?

5. What information could be added to the annotation to improve it?

6. What has the candidate done to improve the flow on their page?

Quick Test

1. **State three** techniques a designer could use to communicate.

2. What is the purpose of annotation?

Purpose of Modelling

Modelling can be valuable at any stage of the design process. Modelling can allow you to:

- generate ideas
- visualise an idea in 3D
- test working or component parts
- solve ergonomic, functional and aesthetic problems.

Any model you make must communicate something you have not shown or discovered in your design work so far.

In industry, models can be used to allow members of the target market to see the concept in a full-size physical form and to provide feedback to the designer.

Materials

You should familiarise yourself with various modelling materials, so that you can use them effectively to communicate and test aspects of your design.

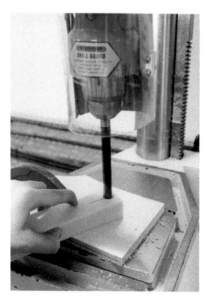

Styrofoam is quick and easy to shape and form. It can be drilled, sanded and finished.

Rather than using scissors, a scalpel and safety ruler have been used to produce a perfectly straight cut.

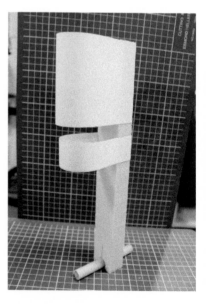

The complete model can be evaluated in terms of scale, proportion and basic functionality.

It is important to remember that marks are not awarded for simply making a model. You must write down what you have learned from the model and it must communicate something that your sketches do not.

Case Study

In the example shown opposite, a candidate has developed a concept for a lamp. The model was produced to evaluate the following aspects of the design:

- stability, to ensure the lamp wouldn't fall over (safety)
- access to the bulb, to ensure there is enough clearance space (ergonomics)
- the look of the assembled model (aesthetics)
- assembly, to figure out where to house the bulb (function/assembly).

This model allowed the candidate to establish the shortest length for the round bar that would still allow the lamp to be stable.

Once the shade was attached, the candidate was able to establish that there was not enough room to access the bulb. This informed a need for further development.

When evaluating the aesthetics of the model, the assembled lamp did not fit within the requirements of the target market as the shade was too bulky to be used on a desk. This informed another strand of development.

Finally, the candidate discovered they could not access the bulb, due to the form and size of the shade. This impacted how the bulb would be attached to the lamp, giving further design problems to resolve.

This particular scale model helped the candidate to gain a better understanding of numerous aspects of their design, which would not have been possible to get from a sketch.

Quick Test

1. **State** where modelling can be used in the design process.

2. **List** the reasons for modelling.

Modelling Techniques

The modelling techniques you should be aware of for the Question Paper and your Course Assignment project are:

- test modelling
- scale modelling
- 3D CAD modelling
- prototyping
- use of mock ups.

Handmade physical models are not the only types of models that you can use. 3D CAD models can also be very useful.

There are good reasons for using CAD modelling:

- Files can be sent to a CNC (Computer Numeric Control) machine for manufacturing.
- Multiple changes can be made quickly and easy.
- Actual materials can be applied digitally.
- It uses no material.
- Multiple copies can be saved and emailed anywhere.
- A workshop environment and equipment is not needed.

CAD models are not always quicker to create than physical models, depending on the complexity of the model and the skills of the designer.

Although software can analyse materials, creating a physical prototype will allow you to carry out 'hands-on' testing and evaluation. Making a physical model is the only method that will enable you to touch and interact with the product, testing how it feels in your hands and experiencing the ergonomics of the shape, form and sizes used.

Quick Test

1. **List** the advantages and disadvantages of CAD modelling.

2. **State five** different types of model.

Idea-Generation Techniques

What Techniques Should I Know About?

There are a number of idea-generation techniques that can inspire creative ideas. These are:

- taking your pencil for a walk
- morphological analysis
- brainstorming
- technology transfer

- analogy/biomimicry
- mood/lifestyle boards
- six-hat thinking
- SCAMPER.

In the Question Paper you will only be asked about brainstorming and morphological analysis. You could be asked to name one or describe the key stages of an idea-generation technique. There are three key stages that apply to most idea-generation techniques:

1. Planning – consider resources and people required, factors/requirement of the task.

2. Conducting – generating ideas, any rules/restrictions such as random selection or no idea discounted.

3. Summary – identify the ideas with the most potential and disregard others.

You also need to be able to generate diverse ideas in Assignment Design and you can use any suitable technique. Practising a range of these will better enable you to choose the correct method depending on the task you are working on.

Practicing and applying some of these techniques during your coursework or in preparation for your Assignment is the best way to develop your understanding of them.

Taking Your Pencil for a Walk

Taking your pencil for a walk is an idea-generation technique that enables unique and interesting shapes to be produced. It is done individually and quickly; very little conscious thought is required.

> **Link to taking-your-pencil-for-a-walk worked example**
> https://collins.co.uk/pages/scottish-curriculum-free-resources

Task

Try taking your pencil for a walk! Simply sketch a rectangle around the inside edge of a piece of paper and plot 30 points randomly around the edge of rectangle. Connect points together in pairs by sketching lines, curves or arcs between them. Next, pick out interesting shapes from the 'grid'. The shape could be used for the front, side or top profile of your design or it could be the shape of one component.

Morphological Analysis

Morphological analysis is suitable for helping individuals who are working on their own to generate a wide range of ideas. It is a very structured and visual process that encourages divergent thinking and problem solving.

The process is fairly straightforward and involves producing a simple table with columns and rows. Once the key **attributes** of the product have been established, they are written as the headings for each column. The next stage is to list a range of different options for each attribute – this fills up the rows.

Link to morphological-analysis worked example
https://collins.co.uk/pages/scottish-curriculum-free-resources

ADVICE

Don't <u>choose</u> attributes from each column. It's better to select attributes at RANDOM.
Try rolling a dice to make each selection completely random. Your results will be unexpected and useful!

Task

Let's try using morphological analysis in an unusual context: we're going to generate ideas for a new creature! This will show you how morphological analysis can be used to generate interesting ideas quickly.

The key attributes for the creature are used as headings at the top of each column and the rows are populated with different options for these attributes.

Colour	Number of eyes	Transport	Limbs	Skin	
WHITE	1	WALK	0	FEATHERS	
YELLOW	2	FLY	2	SHINY	
GREEN	3	SWIM	4	LEATHER	
BLACK	4	SWIM+FLY	6	PALE	
BLUE	5	WALK+SWIM		SCALES	
RED	6	WALK+FLY		FUR	

It is now straightforward to produce ideas for the creature. RANDOMLY select one attribute from each column. For example, if we choose **blue, 5 eyes, swim + fly, 2 limbs and shiny**, we can derive a novel creature.

Brainstorming

Brainstorming can be an individual or group activity. Brainstorming enables designers to *quickly* generate a large range of ideas, working towards a potential solution to a problem. There are no right or wrong suggestions during a brainstorming activity. The aim is to generate a **variety** of ideas, including the weird and wonderful ones.

> **Link to brainstorming worked example**
> https://collins.co.uk/pages/
> scottish-curriculum-free-resources

Technology Transfer

This is the process of **transferring** skills, knowledge, technologies or methods of manufacturing to unrelated products. This technique helps designers develop and exploit technology into new products or applications.

The Dyson vacuum cleaner is a good example of this. James Dyson used cyclone technology to produce a better performing vacuum cleaner. The existing technology came from an air filter in a factory. Dyson was able to transfer this technology to a new application in a different product.

Analogy/Biomimicry

An analogy is simply a **comparison** between two things. A designer can use analogies to generate unusual ideas by finding connections between their design problem and other products or the environment around them.

> **Link to biomimicry worked example**
> https://collins.co.uk/pages/
> scottish-curriculum-free-resources

Biomimicry is similar to analogy, but comparisons are drawn from **nature**.

Lateral Thinking

Lateral thinking is approaching a problem from a different and unconventional perspective. For example, if you were designing a toaster, you might produce a design with two or four slots in the top, as seen in the majority of current models. When **thinking laterally** about toaster design, you might consider slotting the toast in from the front or rotating the toast. Lateral thinking can and should be applied to all the idea-generation techniques – try to think 'outside the box'!

Quick Test

1. What is an advantage of using an idea-generation technique?

2. **List five** idea-generation techniques.

Lifestyle/Mood Boards

Lifestyle boards enable the designer to visualise the target market's interests, needs and preferred styles.

A good lifestyle board allows the viewer to identify the market's age range, gender, style, type of employment, and other interests. A good final design for a product would not look out of place if it were added to the image board!

A **mood board** is different from a lifestyle board. A mood board focuses on communicating the **theme** or **style** of the product, as opposed to communicating the lifestyle and interests of the market.

Images on a mood board show shape, colour, texture, material, environment and attitude. These provide the designer with inspiration for their design.

Task – Idea Generation

Mood Board for Idea-Generation Task

The above image shows an effective mood board that communicates child-like themes. Use this mood board to generate ideas for a storage box for children's toys. You could try out one of the idea-generation techniques that you have been learning about.

Answers on page 141.

Six-Hat Thinking

Six-hat thinking is a group working tool that can be used to generate ideas quickly and evaluate them effectively. It can be used to explore and refine design ideas.

Design team members can learn how to separate thinking into six clear roles. Each thinking role is identified by a coloured symbolic 'thinking hat'. By pretending to wear these hats and assume a role, all participants are forced to think in a different way.

> **Link to six-hat thinking worked example**
> https://collins.co.uk/pages/scottish-curriculum-free-resources

 The white hat calls for the information and facts that are needed.

 The yellow hat symbolises brightness and optimism; this person explores the positive aspects of an idea.

 The black hat is all about using judgment – picking out flaws in an idea and considering why an idea may not work.

 The red hat signifies feelings and intuition. The wearer of this hat can express emotions and feelings, sharing likes, dislikes, loves and hates.

 The green hat focuses on creativity and this person explores the possibilities surrounding an idea. New ideas can develop from existing ones.

 The blue hat wearer manages the entire thinking process, coordinating all the other hat wearers.

S.C.A.M.P.E.R

This tool helps generate ideas for new products by encouraging a focus on improving existing ones.

> **Link to S.C.A.M.P.E.R worked example**
> https://collins.co.uk/pages/scottish-curriculum-free-resources

S – Substitute

C – Combine

A – Adapt

M – Modify

P – Put to another use

E – Eliminate

R – Reverse

Applying these actions to an existing product or idea can prove really useful when developing or exploring an idea.

Development

You should spend the majority of your design time doing *development* work. In the Assignment, it is where most of the marks are awarded, as development allows you to demonstrate knowledge and skills.

The development stage is your opportunity to **explore**, **refine** and **communicate**.

The key to good development is ensuring that there is purpose. There is no 'one-size-fits-all' process in terms of developing a product. Development is not about changing the *look* of your initial design. It is about working through a meaningful process that leads to the best solution.

Throughout the development stage you should:

- **identify** the problems that you have to solve for your *unique* design task
- **identify** what facts you need to find out to ensure you can develop the best solution
- **explore** alternatives to ensure you create the best solution
- **refine** the solution, adding sufficient detail to permit manufacture.

A successful development will apply knowledge and understanding of the **design factors** and **materials and manufacturing**.

Starting Development

You should have chosen an idea or aspects of different ideas, based on their strengths. Try not to be too precious about the idea at this stage, as it is only a staring point. Remain open to the belief you can improve it. If you thoroughly explore a variety of alternatives, reasoned change will occur naturally to improve some, if not all, aspects of your design. If you find exploration difficult, you could try creating a development plan to keep you focused throughout this stage. You could use a basic graphic organiser like the one shown here. Your plan should not be worked through in order like a list. Instead allow the development to take a natural course and explore and refer to the things on your plan when appropriate.

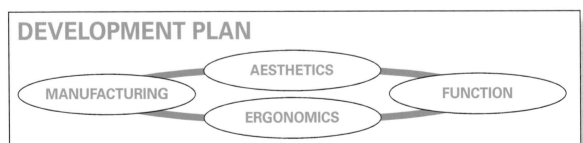

DEVELOPMENT PLAN

AESTHETICS

MANUFACTURING

FUNCTION

ERGONOMICS

Consider alternative ways to hold – in, on, hanging. Alternative arrangement – angled, stacked, side-by-side.
Do the aesthetics suit the theme/market? Consider proportion, balance, contrast, symmetry, materials etc.
Is it easy to use? Consider alternative sizes, ways to interact and use, modelling and testing.
Does it allow me to demonstrate: marking out, use of a range of tools and machines, assembly? Alternatives for joining- forming- shaping-detail? Can it be made in the time?

Case Study

In this example, the candidate has made use of a development plan for a desk lamp project. They have clearly identified the need to explore areas of aesthetics, function and ergonomics.

> **Link to lamp design work**
> https://collins.co.uk/pages/
> scottish-curriculum-free-resources

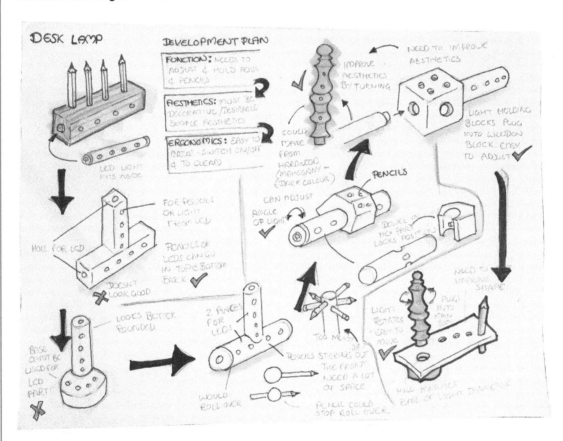

The candidate has explored alternative locations for the light fitting, how the lamp can adjust and where the pencils will be stored.

The arrows help communicate the flow of the design thinking. The ticks and crosses help communicate where the decisions are being made.

The candidate has successfully explored a variety of solutions for various aspects of the design.

Assignment Tip

For development to be successful, you must explore a range of alternative solutions for a variety of aspects within the design.

Refinement

Once you have explored all of your options, your design will have evolved and become more detailed. To refine the design, you need to finalise all the little details to ensure your product meets the **specification** and is suitable for manufacture.

Research is an important **problem-solving** tool for designers at all stages of the design process, and you will certainly have carried out research during your design journey. Whether producing ideas, developing or refining a design, make sure your research is meaningful and use it to draw conclusions that inform your work. Don't just archive information that you found. If the information is not relevant, do not use up valuable space by including it. If the information is relevant, apply it in the development of your product. To demonstrate your understanding of how to use research properly, you also need to show how your results led to a new idea, development or refinement.

Modelling is an essential tool for refining ideas as it enables you to test various aspects of your design, from working out sizes and proportion to testing stability or working parts. Having a physical model is more meaningful than justifying changes through sketch. Because the model is 3D, it allows you to develop a greater understanding about your idea. However you decide to refine your idea, you need to record your decisions through written annotations.

When refining the design:

- check your solution meets all requirements of the specification
- finalise and communicate all sizes so that manufacture can take place
- make and record informed decisions about the materials you require
- make and record informed decisions about the assembly methods.

Assignment Tip

You will need to create a **working** drawing or dimensioned sketch for your assessor to check the accuracy of your practical work. When you carry out refinement in the folio pages to work out sizes and assembly, you will be awarded marks for knowledge of materials and manufacturing. You must however transfer this information onto the planning for manufacture proforma to be awarded any of the marks for planning.

Case Study

In the example below, the candidate has refined necessary areas of the design. This is essential to ensure that the final product is functional and that the design is suitably detailed to permit manufacture of the prototype.

Link to refinement worked example
https://collins.co.uk/pages/scottish-curriculum-free-resources

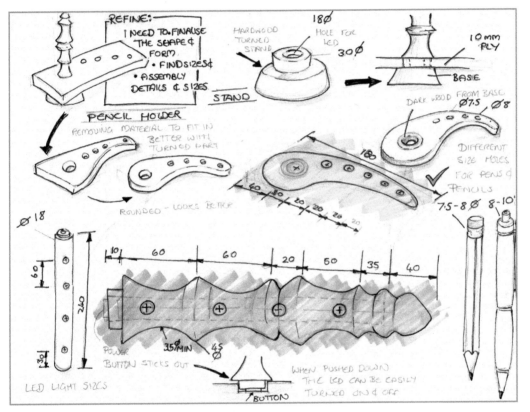

The candidate has made final decisions on the **aesthetics** of the product. **Research** has been carried out and applied to ensure the product is functional.

Detailed, dimensioned sketches have been produced to be used as **working drawings** for each component. These provide clarity and give the necessary sizes required to manufacture and assemble the product.

Quick Test

1.   What decisions has the candidate made in the example above?

2.   **List** the things that should be done when refining a design.

3.   How could you use research to refine or develop your design?

Evaluation

Designers need to be very good at evaluation. Remember, in industry, designers are developing products for markets, not for themselves. This means they could be designing products that they do not like! Continually evaluating and reviewing design decisions against the specification keeps the designer on track for developing a successful solution that meets the needs of the **target market**.

Evaluation can be used at various stages throughout the design process. The **specification** should be referred to as a guideline to **benchmark** the success of the ideas for the final design.

Evaluating and Decision Making

When developing your design solution, you will need to explore a range of alternatives. By doing this you will evaluate and apply knowledge of design and/or materials and manufacturing to make informed decisions. It will be clear that you have applied knowledge and evaluated if you have compared:

- the strengths in an idea
- the areas that require development.

Consider some approaches that could be used to evaluate an idea: star, traffic-light and web rating.

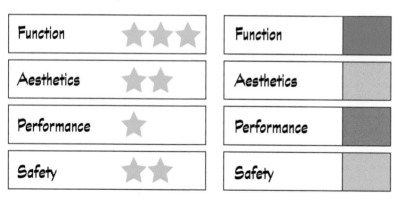 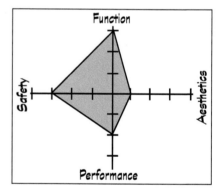

Method A: Star rating   **Method B: Traffic-light rating**   **Method C: Web rating**

All of these methods highlight areas of strength and areas for development. However, used alone, they fail to communicate any real understanding; without any explanation, **these methods are meaningless**. These methods are unnecessary and also time consuming and demonstrate no understanding or reasoning for the decisions/scores.

Simple annotations that explain the problems or strengths of an idea are all that are required during the design process.

A good evaluation helps the designer make decisions about how to move forward in the development process. An example of evaluative annotations is shown below.

Function	This design is suitable for evening reading because the shade prevents direct bright light. The bolts on the neck make the lamp easy to adjust.
Performance	The bulb is not currently easy to access. This would make it difficult to replace and maintain the lamp. Further exploration is required.
Safety	This design uses a low-energy bulb, which should prevent overheating and fire. The design looks like it might be unstable. If it falls over it could break or damage its surroundings. A more stable base is necessary.

Evaluating the Proposal

At the end of Assignment Practical, you need to evaluate the success of your manufactured proposal. If the designer has done a good job, they will be able to justify how the solution meets all of the specification points, or suggest what further changes need to be made.

You should apply research and evaluation techniques, such as testing that it carries out its function, surveying the target market to get opinions on the appearance, or asking them to carry out a user trial to feed back on the ease of use. It is important that your evaluation is more than your personal opinion.

When you evaluate your product, you should make reference to the points in your specification to see how well you met the criteria.

You are not required to comment on your ability during the design process or your practical skills.

Assignment Tip

Evaluating throughout your development work will attract marks. When evaluating, ensure you refer to all aspects of the specification. You will be assessed on the depth of your evaluation and the appropriateness of the techniques used. Your evaluation must be based on more than personal opinion.

Quick Test

1. Why are star ratings, on their own, not a suitable evaluation method?

2. What should the designer refer to when evaluating a design idea?

3. How can evaluating help the designer make decisions?

The Proposal

Once you have refined the details of your design and manufactured your prototype, you may wish to produce a **design proposal**. In industry, designers produce design proposals to allow the client (and prospective buyers) to **visualise** and understand the benefits of their design.

Producing a proposal using photographs of your Assignment Practical will not attract any marks in the N5 Assignment; however, it is a good way to communicate aspects of your final design during the evaluation.

If your proposal is paper based, it can be helpful to communicate the scale of your design by showing it in context. Compare the candidate examples shown below.

Proposal 1

Proposal 2

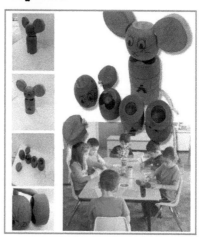

The first proposal communicates very little about the product. It provides limited information about the scale or function of the design.

The second proposal provides us with far more information. Showing the product in use gives us an idea of the **scale** and **market**. The different views of the design demonstrate additional **functions** of the product.

Annotating the proposal to highlight the key features and selling points would further enhance the viewer's understanding and interest in the product.

Planning for Manufacture Revision Questions

In this section you will find questions that cover planning for manufacture. Use these questions to test your current knowledge and understanding, and to identify areas for further revision. You should be able to answer all of these questions by the time you are ready to sit the Question Paper. Answers are given on page 135.

You can find additional information on this topic on these pages:

- **Working Drawings (WD)** – Page 88
- **Exploded and Detailed Views (V)** – Page 90
- **Cutting Lists (CL)** – Page 91
- **Sequence of Operations (SO)** – Page 92
- **Evaluating the Plan for Manufacture (E)** – Page 93

1. **List three** pieces of information you could find on a working drawing. (WD)

2. At what stage in the design process would you produce a working drawing? (WD)

3. **Explain** how a working drawing can help during the manufacturing of a model or prototype. (WD)

4. What are the benefits of using exploded or detailed views? (V)

5. What information must be included in a cutting list? (CL)

6. **Explain** what could happen during the manufacturing of a model, if a detailed sequence of operations is not produced. (SO)

7. What **three** components (parts of the plan) should be considered when evaluating a plan for manufacture? (E)

Working Drawings

In your Area of Study work and your Course Assignment, you will be required to produce a **working drawing** that details all of the information required to manufacture a product. The following task will allow you to see the potential problems that you might face if your working drawing does not include the correct information.

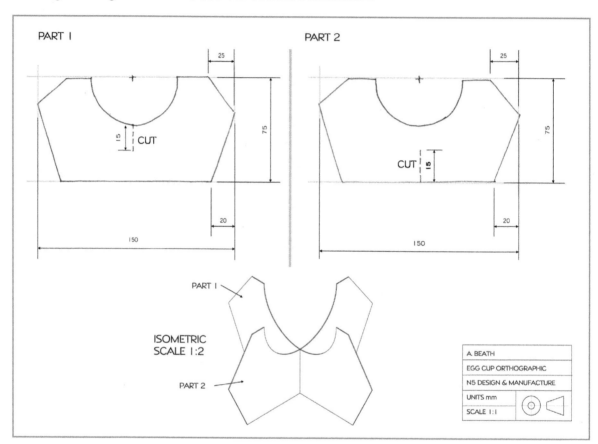

Task

Try to make an accurate model of an egg cup using the information on the working drawing shown above. The most appropriate modelling material for this task is card.

You will find that it is impossible to complete the model as you are not given enough information.

List the missing information that would allow you to complete the model. Alternatively, you could add the information to the working drawing itself.

Answers are on page 141.

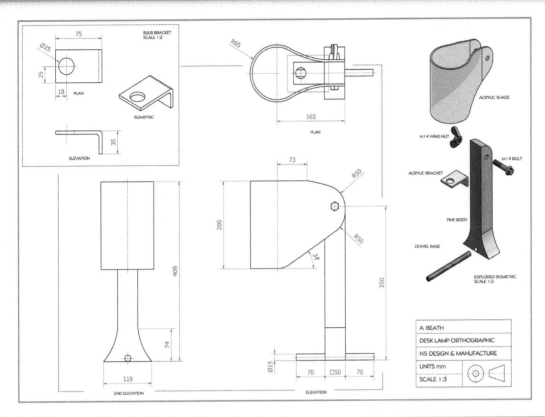

This is a candidate's final working drawing for a lamp. It has been produced using **CAD** software.

The information on this document will enable the candidate to manufacture the components of the lamp and assemble them accurately.

A good working drawing, like this example, should include:

- relevant orthographic views (elevation, end elevation and plan)
- drawings showing all component parts of the product
- all necessary dimensions to permit manufacture
- any technical detail required to permit manufacture.

Link to CAD working drawing of a lamp
https://collins.co.uk/pages/scottish-curriculum-free-resources

Assessment Tip

You need to produce a dimensioned drawing or sketch for your planning for manufacture pro forma in the Assignment. It should be clear from the graphics what the product is, and parts, sizes and assembly details should be communicated.

Quick Test

1. **Explain** why a poor quality working drawing can cause problems for a manufacturer.

2. What types of information can be found on a working drawing?

Exploded and Detailed Views

As you develop and refine your ideas, you will begin to understand the different parts that are needed to make up your design and how they fit together.

Exploded Views

In the example shown on the right, the mouse design has been developed and is almost ready to manufacture. Due to its complexity, it is difficult to see how all of the different parts fit together. The designer modelled an exploded view of the mouse to show this more clearly.

You can create exploded views in a number of different ways:

- sketching your design with all of the parts separated
- creating an exploded view using CAD software
- creating a physical model and separating the parts.

When refining assembly details for a computer mouse, for example, you could look at existing mouse designs and how they are assembled. This would help you generate exploded views of your own design.

Detailed Views

You are likely to sketch and model ideas that have specific details or features, such as joining components and moving parts. A good way to show these parts is to create detailed views. These show particular parts on a larger scale.

You can create detailed views in a number of different ways:

- sketching parts at a larger scale
- taking close-up photographs of physical models
- zooming in close to a rendered 3D model.

Cutting Lists

A **cutting list** is a list of all the parts that make up your prototype. The information on the cutting list will help you to plan and prepare your materials. The list, which is normally written in the form of a table, should contain the following information:

- the length, breadth and thickness of each part
- how many of each part you require
- the material required.

It is important that the sizes stated in the cutting list are taken from your **working drawing**. It is a good idea to give each part a number or name. This will help you to identify the parts more easily, especially if you are manufacturing a prototype with numerous components.

The cutting list for the lamp from page 89 is shown here. You may notice that both of the acrylic parts are listed in their original flat form, before they are heated and bent into shape.

Part	No.	Material	Length	Width	Thickness	
Main Body	1	Pine	400	119	50	
Dowel	1	Birch	190	Ø15		
Bracket	1	Black Acrylic	110	50	3	
Shade	1	Clear Acrylic	600	200	3	

Assignment Tip

You will need to complete a cutting list as part of the planning for manufacture pro forma. Ensure the sizes or parts in your cutting list are also easy to identify in your supporting drawings.

Quick Test

1. **List** all the information found in a cutting list.

2. What **three** methods could you use to create an exploded view?

Sequence of Operations

Preparing for manufacture is a very important stage. Failing to organise the stages of manufacture into a logical order could result in an inaccurate and flawed model.

This part of the manufacturing plan is known as a **sequence of operations**.

When you are preparing your sequence of operations you should consider:

- the logical order in which to carry out the manufacturing steps
- the tools you will need to use for marking, cutting and shaping
- preparation of materials and application of a finish
- how the different parts fit together and the process of assembly.

Your sequence of operations should contain as much detail as necessary so you or anyone else can follow it as a guide to manufacture the product. Your sequence of operations *must* be completed *before* you start making your final model.

Your sequence of operations should link closely to your **working drawing**, which will have all of the key **dimensions** for your design.

Assignment Tip

In your Assignment, your sequence of operations is split into steps and tools. You are given limited space to complete it. The tools and equipment must be linked to the stage to attract marks.

SEQUENCE OF OPERATIONS FOR TURNING A PEN

1. Mark the diagonals on the ends of the blank.
2. Use a steel rule and try-square to measure and mark the length of each half.
3. Using a tenon saw, cut to size, then sand.
4. Use the pillar drill to create a 7-mm hole through the centre of the diagonals.
5. Glue the barrels in the blank and allow to cure.
6. Mount the blanks onto the mandrel on lathe.
7. Use a scraper to remove the flats.
8. Mark the position of detail using a pencil and steel rule.
9. Use a gouge and skew chisel to shape the pen.
10. Finish using different grades of glass paper before applying wax using a cloth.
11. Remove from lathe; insert nib first.
12. Carefully assemble the remaining components.

STAGE	TOOLS
1. Mark the blank	steel rule/ try-square
2. Cut and drill the blank	tenon saw, pillar drill, 7-mm twist drill

Quick Test

1. At what stage in the design process would you produce a sequence of operations

2. **List three** pieces of information you would find in a sequence of operations.

Evaluate the Plan for Manufacture

Task

For this task, you are going to try to produce a card model of a flat-pack chair that is made up of four parts. You should use the following information:

- working drawing
- cutting list
- sequence of operations.

Unfortunately, some information is incorrect or has been missed out completely. Your challenge is to identify:

1. incorrect or missing information from the **working drawing**
2. missing information or inaccuracies in the **cutting list**
3. missing information from the **sequence of operations**
4. any **tools** that were missing from the sequence of operations.

Answers are on page 141.

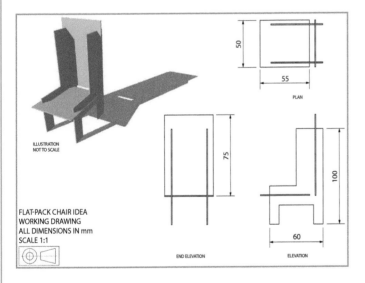

ILLUSTRATION
NOT TO SCALE

FLAT-PACK CHAIR IDEA
WORKING DRAWING
ALL DIMENSIONS IN mm
SCALE 1:1

50

55

PLAN

75

100

60

END ELEVATION   ELEVATION

Cutting List

Part	No.	Material	Length	Width	
Sides	2	Card	100	60	
Seat	1	Card	55	20	
Back	1	Card			

Tools required for manufacture:

Pencil
Ruler

Sequence of Operations:
– Sides (x2)
1. Draw a rectangle 100x60.
2. Draw the position of legs.
3. Using scissors, cut out the shape.
4. Cut both the slots for the seat and back to fit into.
5. Repeat this process and then colour both parts red.

– Seat
1. Draw a rectangle 50x55.
2. Cut out the shape.
3. Colour the rectangle green.

– Back
1. Draw a rectangle.
2. Cut it out.
3. Colour the rectangle green.

– Assembly
1. Slot the seat into the bottom slot.
2. Slot the back into the top slot.

Workshop Manufacture Revision Questions

In this section you will find questions that cover Workshop Manufacture. Use these questions to test your current knowledge and understanding, and to identify areas for further revision. You should be able to answer all of these questions by the time you are ready to sit the Question Paper. Answers are on page 135.

The following page numbers show where you can find additional information on this topic:

- **Measuring and marking out (M)** – Page 96
- **Cutting, shaping and forming (C)** – Page 98
- **Finishing (F)** – Page 100
- **Assembling (A)** – Page 102
- **Health and safety (HS)** – Page 104

1. **State** the name of **three** tools used to mark out wood. (M)

2. **State** the name of **three** tools used to mark out metal. (M)

3. How can you check the accuracy of your marking out? (M)

4. Other than saws, what tools are considered cutting tools? (C)

5. **Explain** the difference between shaping and forming. (C)

6. **State two** pieces of equipment used to form plastic in the workshop. (C)

7. **List** the steps used to prepare the following materials for finish (F):

   a) wood

   b) aluminium

   c) acrylic.

8. **List three** finishes that can be applied to (F):

   a) wood

   b) metal.

9. **State** the advantages of using jigs and fixtures during assembly. (A)

10. **List six** ways to join and assemble materials. (A)

11. **List five** types of protective clothing and equipment in the workshop. (HS)

12. What **two** wood work joints have been used in the manufacture of the infant's toy shown above? (A)

13. What tools would be used to mark out the corner joints on the frame? (M)

14. What tools would be used to cut the corner joints on the frame? (C)

15. What methods can be used to ensure the frame is glued squarely? (A)

16. What finish has been used to produce the green colour on the base of the toy? (F)

17. What **two** machines could have been used to form the plastic parts of the toothbrush holder shown above? (C)

18. How can you prevent plastic from cracking, while drilling? (C)

Measuring and Marking Out

Measuring and marking out your material is another important stage in the manufacturing process. A common mistake made by pupils is rushing the measuring and marking out. It is important to take your time. The more careful you are when marking out, the more accurate your cutting and assembly will be.

Remember, you are assessed on accuracy, not on how quickly you make your product.

When parts are accurately marked out, the cutting, shaping and assembly tends to be more straightforward. You should maintain a careful approach throughout the manufacturing stages to ensure your work remains accurate, neat and well finished.

Task – Wood

There is a wide range of marking-out tools for wood. Although you are unlikely to use them all, you still need to know about them as you may need to refer to them in the Question Paper.

Some common marking-out tools are shown in the table below. **Complete** the table, giving a precise definition for each tool. One has been done for you.

You should carry out some research to help you identify the correct definitions. Answers are on page 142.

Tool	What is it used for?	
Steel rule		
Marking gauge	Marking a line parallel to an edge	
Try-square		
Mortise gauge		

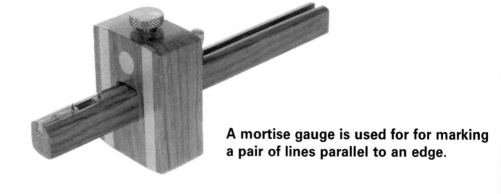

A mortise gauge is used for for marking a pair of lines parallel to an edge.

Task – Metal

There are special tools for marking out metal. You may need to familiarise yourself with these for the Question Paper.

Some common marking-out tools for metal are shown in the table below. **Complete** the table giving a precise definition for each tool. One has been done for you.

You should carry out some research to help you identify the correct definitions. Answers are on page 142.

Tool	What is it used for?
Scriber	Marking lines that won't rub off onto metal
Odd-leg callipers	
Engineers square	
Spring dividers	
Centre punch	

Marking Out Plastic

Many woodwork and metalwork tools can be used for marking out plastic. There are no tools that are only used for marking out plastic. Normally, a water-based pen (as opposed to a pencil or scriber) is used to mark out lines. This wipes off easily, leaving no evidence of marking out.

Acrylic has a sticky film coating that protects it from being scratched. You can mark lines directly onto the film and peel it off once you have finished cutting, shaping and finishing the acrylic.

Quick Test

1.   Why is marking out accurately important in the manufacturing process?

Cutting and Shaping

Once you have marked out your material, you will be ready to start cutting and shaping. The type of cutting or shaping you are carrying out, and the material you are working with, will determine the tools you should use.

Selecting the correct tools can be tricky. For example, you are probably familiar with three or four different saws. So, which one do you use?

Task 1

The table lists tools and processes that are suitable for different materials.

Complete the table, stating the tools and materials that apply to each process. Remember, there may be more than one material per process. Answers on page 142.

Tool	Material(s)	Process
		Removing material for a mortise hole
		Cutting intricate curves and shapes
	Wood	Cutting straight lines, often across the grain of the material
		Making a rip cut, parallel with the grain of the material
		Cutting larger materials – can handle curves and straight lines (electric hand tool)
		Carving or cutting – ideal for removing thin pieces of material at a time
	Metal	Ideal for cutting straight lines – has replaceable blades
	Metal	Cutting sheet material
		Cutting fine amounts of material away, in straight lines, internal curves and external curves
		Shaping and smoothing the end grain of the material

Task 2

Look at the prototype of the clock shown below and answer the questions.

Answers on page 142.

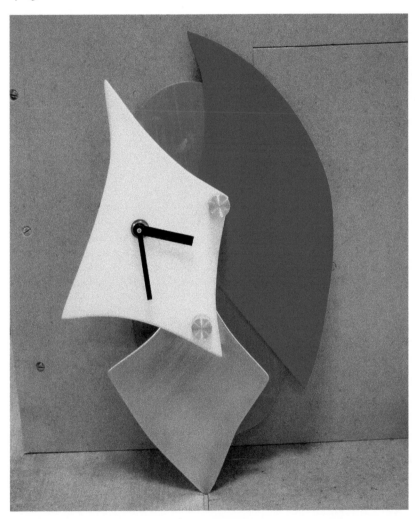

1. What machine could have been used to manufacture the **two** aluminium spacers.

2. What is the name of the process that is carried out on the machine that produces the attractive surface finish on the ends of the aluminium bars?

3. **Describe** how the candidate could have cut and shaped the:

    a) sheet aluminium

    b) white plastic.

4. **Name two** finishes that could have been applied to give the wood the blue colour.

Finishing

In the Area of Study work and Course Assignment, you will be assessed on the quality of **finish** you apply to the different parts you manufacture in the workshop. In the excitement of completing a project, many candidates rush the finishing stage. You should regard finishing as just as important as the other manufacturing stages, taking your time to give your product the best possible finish.

Wood

Before applying a finish, you should prepare wood carefully. Glass paper should be used with a sanding cork to remove cutting lines, glue spots and pencil marks. Start with a coarse grade of glass paper, such as P60, and finish with a fine paper, like wet and dry.

Here are some of the types of wood finish you should know about:

- stain and dye
- paint
- lacquer
- varnish
- wax
- oil.

Metal

Metal is a much harder material than wood but, by going through the correct finishing process, you will still be able to achieve an excellent finish. Just as with wood, to get the best results, you should prepare your metal parts properly before applying a finish.

Task 1

Complete the table below by identifying the different stages and processes involved in preparing sheet aluminium for a finish. Answers are on page 142.

	Step	Reason	
1	Filing	Removes burrs and sharp edges	
2	Emery cloth		
3	Wet and dry		
4	Polish		

Plastic

Plastics, such as acrylic, tend to be a lot easier to finish than wood or metal. It is fairly straightforward to achieve an excellent finish on the edges, providing you follow the correct process and take your time.

The process of finishing the edges of plastic is something that you could also be asked about in the Question Paper.

Task 2

The table shows a part-completed sequence of operations for finishing the edge of acrylic. The first stage has been completed. Finish the rest of the sequence, making sure you explain the purpose of each stage. Answers are on page 142.

	Step	Purpose
1	Cross file	This should remove all of the saw marks. It will leave scratch lines across the material.
2	Draw file	
3		
4		

Task 3

Create a 'mind map' with the heading 'Finishes'. Make branches for each of the three material types (wood, metal and plastic) and then identify the finishing processes for each. Answers are on page 142.

Quick Test

1. **List six** different finishes that can be applied to wood.

2. **Describe** the processes of finishing wood using glass paper.

Assembling

The last stage of your manufacturing process will be the **assembly**. How well your parts assemble will be determined by the accuracy of your marking out, cutting and shaping. If you have worked with care and accuracy, your parts should fit together nicely.

Depending on the materials you use, there will be a range of techniques for assembling the different components. During the design and development phases, you should have gained a fairly clear idea of how you will assemble your model. This should be shown in your sequence of operations.

Material can be assembled in a range of ways. You must decide, during your design development, which methods are the most suitable. Some examples are shown in the boxes below.

Wood	**Metal**	**Plastic**
• wood joints	• tapping and threading	• solvent cement
• wood screws	• riveting	• pop rivets
• knock-down fittings	• nuts and bolts	• screws (onto wood)
• adhesives	• welding screws	• adhesives
• nails		
• hinges		

Knock-Down Fittings

Knock-down fittings are **standard components** that are used to temporarily or permanently fix materials together. They are common in modern **flat-pack** furniture that consumers can build at home.

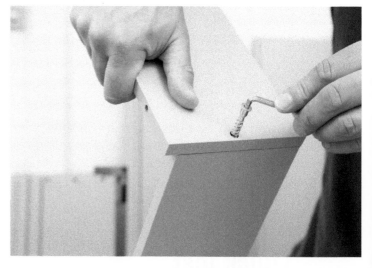

Both manufacturers and consumers benefit from knock-down fittings. Manufacturers buy them in vast quantities, reducing the cost of each fixing. This cost is recouped on each piece of furniture that is sold. Also manufacturers do not need to assemble the furniture in their own factories, which saves time and money. As a result they can release products into the market more quickly at lower retail cost. This results in the products being more affordable to a larger proportion of the available market.

Case Study

IKEA make great use of knock-down fittings in their furniture. All of their products, including wardrobes, shelving units and bed frames, are flat-packed, meaning that the consumer can transport the furniture home in their own cars and not incur expensive delivery charges.

IKEA use laminated manufactured boards, such as MDF and chipboard, in their furniture. This also reduces the costs for the consumer.

Building the furniture is simple as each piece comes with illustrated step-by-step instructions. Only simple tools are required.

Task

Carry out research and identify as many knock-down fittings as you can find.

You should **list** each one and **state** where they are commonly used. An example is shown below.

Knock-down fitting: plastic corner block

Typical use: kitchen cabinet carcasses

Answers are on page 143.

Quick Test

1. **State** a benefit for manufacturers of knock-down fittings.

2. **State** a benefit for consumers of knock-down fittings.

Health and Safety

Health and safety is paramount, both in the workshop and in industry. You should have an awareness of general procedures and the implications of not following them, for both environments.

Task 1

Within the workshop, you *must* demonstrate the ability to work safely, recognising when tools and machines are in safe working order. Wearing eye protection is not always the only or most appropriate precaution. You should ensure you have an understanding of the specific risks related to the tools or machines you are working with and be able to follow appropriate procedures to enable you to carry out your task safely.

Look at the machines shown below. **Complete** the table, identifying **two safety checks** and **two precautions** you would follow to ensure each process was carried out safely and effectively. Answers are on page 143.

Machine tool	Safety checks	Safety precautions	
Pillar drill	1. Remove chuck key	1. Wear eye protection	
	2. Ensure workpiece is secure	2. Wear an apron	
Centre lathe			
Forge			
Band facer			

Health and Safety in Industry

Health and Safety is a serious concern in industry, and companies must ensure they carry out the correct procedures to adhere to Health and Safety regulations. Employers are *legally bound* to ensure:

- they offer a safe environment for their employees
- employees have access to and wear personal protection equipment

- machines are regularly maintained
- staff receive adequate training and are made aware of the risks of their job.

Companies (and schools) routinely carry out **risk assessments**. The purpose is to identify the likelihood and potential severity of harmful events.

Task 2

Create a 'mind map' to explore the different facilities and procedures that might be in place to protect the Heath and Safety of employees. As part of the 'mind map', state briefly why each would be required. Answers on page 143.

Here are a few to start you off.

- First aid – trained first-aiders are always present in a workplace.
- Liability – companies are responsible for the wellbeing of their employees.
- Training – all staff should be trained in Health and Safety.
- Signage/alarms – signage and alarms are used to alert employees to dangers.

Safe Design

Everyone in the design team plays a part in ensuring the product is safe to use and presents no risk to the consumer:

- The designer must consider how the product will be used, to ensure the product will not fail or cause injury.
- Products are put through thorough tests and, to be suitable for retail, must meet regulations like British Standards.
- Any safety warnings must be clearly displayed on the product.
- Quality assurance checks are carried out by the manufacturer, to ensure the reliability and quality of the manufactured product before it leaves the factory.
- Even at the point of retail, if a fault is identified with a product, companies will recall products to protect their customers and their reputation.
- Advances in technology have improved product testing, using simulation and test rigs, and reducing the risk to users.

Question Paper Tip

You may be asked about the safety of a product in the exam. Ensure you respond with an appropriate answer. Think carefully about the product you are shown and the risks or dangers associated with it.

Quick Test

1. **State two** ways employers ensure the safety of their employees.

2. **State two** methods that are used to ensure a product is safe for retail.

Commercial Manufacture Revision Questions

In this section you will find questions that cover planning for manufacture. Use these questions to test your current knowledge and understanding, and to identify areas for further revision. You should be able to answer all of these questions by the time you are ready to sit the Question Paper. Answers are on page 136.

The following page numbers show where you can find additional information on this topic:

- **Understanding material properties** – Page 108
- **Metal (M)** – Page 110
- **Wood (W)** – Page 111
- **Manufactured boards (MB)** – Page 111
- **Plastic (P)** – Page 112
- **Fixing and joining techniques (F)** – Page 114
- **Manufacturing processes (MP)** – Page 115

1. In the table below, **list three** softwoods and **five** hardwoods. (W)

Softwoods	Hardwoods	
	Walnut	

2. **Describe** the environmental benefits of using softwoods, as opposed to hardwoods.(W)

3. **State** the names of **three** manufactured boards. (MB)

4. **Describe three** benefits of using manufactured boards in flat-pack furniture. (MB)

5. **State** the name of a suitable hardwood for a kitchen utensil. (W)

6. **Explain** why red cedar is an appropriate material for a garden shed. (F)

7. **State** the name given to metals that contain **iron**. (M)

8. **Explain** why aluminium is a suitable material for a soft-drinks can. (M)

9. What **processes** can be used to form plastic and metal tubes? (MP)

10. Give **two** reasons why stainless steel would be a suitable material for cutlery. (M)

11. **Explain** the difference between a **thermoplastic** and a **thermoset plastic**. (P)

12. In the table below, list as many examples of **thermoplastic and thermoset plastics** as you can. You can carry out online research to help you with this task. (P)

Thermoplastics	Thermoset plastics	
ABS		

13. **State** the names of **three** plastic commercial manufacturing processes. (MP)

14. **Explain** why **HDPE** would be a suitable material choice for a sports-drink bottle. (P)

15. **Explain** why die casting is a suitable process for manufacturing alloy wheels. (MP)

Understanding Material Properties

You are expected to know about a number of materials in Design and Manufacture. From **woods** to **plastics** to **metal**, materials can become very confusing and it is easy to mix up their properties. The next few pages will show you products, materials and some material properties. Think about why the products are made from particular materials, and which of the materials' properties are important in each case.

You may come across tables with lists of materials and their properties. Trying to memorise every one would be extremely difficult and also a bit boring! One of the best ways to understand material properties is by examining existing products and considering why each material was chosen for the different components, as in the task on the opposite page.

ADVICE

Avoid memorising tables of material properties. Identify what a product MUST DO in order to understand the likely properties of the materials. Sometimes more than one material can be suitable for a product.

Plastipedia

The British Plastics Foundation has an excellent free online resource that you may want to use for your revision for plastics. Follow the weblink on the right or simply use an internet search engine (enter 'Plastipedia' and you will find the website). From here, there are links to processes, properties of polymers and much more. Please note that plastics are often referred to as **polymers** on the site.

Link to Plastipedia
www.bpf.co.uk/Plastipedia/
Polymers/Default.aspx

The commercial manufacturing processes you should know about are:

- injection moulding
- rotational moulding
- vacuum forming.

BBC Bitesize

The National 5 Design and Manufacture section on BBC Bitesize has information on materials and manufacturing processes. The weblink will direct you to these parts of the Bitesize website. Again, carrying out a simple internet search will also take you to the site.

Link to BBC materials
www.bbc.co.uk/education/
topics/zyvrwmn

Be sure to watch the videos that cover materials and manufacturing processes.

Task

Consider the five products shown below. List the properties you think the product should have and identify the appropriate materials. Refer to pages 110, 111 and 113 to help you.

The first example has been done for you. Answers on page 143.

Stool seat
The material is shiny and bright red in colour. It would have to be rigid for people to sit on it safely. Stools are moved around a lot and can also be knocked over, so this material would have to be impact resistant. Therefore the material is likely to be:
ABS
OR
Polypropylene

Stool base
This material is shiny and reflective with a silver colour. It would have to be strong to hold a person's weight. This material is likely to be:
chrome-plated mild steel

1

Cleaning bucket

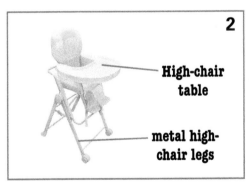

2

High-chair table

metal high-chair legs

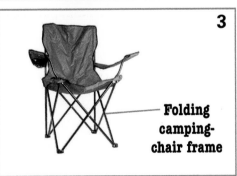

3

Folding camping-chair frame

4

Kitchen-knife handle

Kitchen-knife blade

Materials

Metal

There are two main types of metal you should know about for your exam:

- **ferrous** – (containing iron)
- **non-ferrous** – (without iron).

Metals can be mixed together to form new metals. These are known as **alloys**.

COPPER — NON-FERROUS

WHY IT'S USED
Ductile
Thermal conductor
Electrical conductor
Resistant to corrosion

USES
Electrical wiring
Jewellery
Plumbing pipes

ALUMINIUM — NON-FERROUS

WHY IT'S USED
Soft
Easy to work
Easy to cast
Very good strength to weight ratio

USES
Drinks cans
Saucepans
Ladders
Window frames

CAST IRON — FERROUS ALLOY

WHY IT'S USED
Strong in compression
Relatively inexpensive

USES
Car engine parts
Exterior furnishings such as fences and benches
Manhole covers on roads

HIGH CARBON STEEL — FERROUS ALLOY

WHY IT'S USED
Good wear resistance
Can be hardened and tempered

USES
Cutting tools, files, saws and drill bits
Hammers

MILD STEEL — FERROUS ALLOY

WHY IT'S USED
Malleable
Ductile
Easy to weld

USES
Car body panels
Nuts and bolts
Springs

BRASS — NON-FERROUS ALLOY

WHY IT'S USED
Good corrosion resistance
Hard
Good wear resistance

USES
Screws
Hinges
Door handles
Musical instruments

STAINLESS STEEL — FERROUS ALLOY

WHY IT'S USED
Corrosion resistance
Strong
Hard

USES
Kitchen appliances
Kitchen sinks
Furniture components and accessories (e.g. locks and chair legs)

Wood

There are two types of wood you should have knowledge of for your exam:

- **hardwoods**
- **softwoods**.

SCOTS PINE

SOFTWOOD

WHY IT'S USED
Cream to pale brown in colour
Straight-grained
Quite strong
Easy to work

USES
Furniture
Joinery
General construction

RED CEDAR

SOFTWOOD

WHY IT'S USED
Red-brown colour
Lightweight
Soft and weak
Durable to weather

USES
Garden sheds
Garages
Cladding

OAK

HARDWOOD

WHY IT'S USED
Strong
Durable
Expensive

USES
Furniture
Flooring

BEECH

HARDWOOD

WHY IT'S USED
Pale brown colour
Food safe
Strong

USES
Kitchen utensils
Furniture
Turnery

ASH

HARDWOOD

WHY IT'S USED
Light colour
Tough
Flexible

USES
Sports equipment
Tool handles

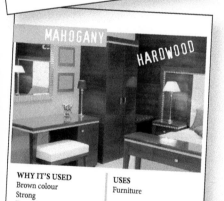

MAHOGANY

HARDWOOD

WHY IT'S USED
Brown colour
Strong
Durable
Easy to work

USES
Furniture

MDF

MANUFACTURED

WHY IT'S USED
Flat, paintable surface
Stiff
Machined easily

USES
Indoor furniture
Radiator covers

PLYWOOD

WHY IT'S USED
Uniform strength across the board
Flat/stable
Resistant to warping and twisting

USES
Indoor furniture
Wall panelling

Manufactured Boards

You should familiarise yourself with the different manufactured boards, their properties and uses.

CHIPBOARD

MANUFACTURED

WHY IT'S USED
Low cost
Stable
Usually veneered

USES
Indoor furniture
Laminated for kitchen worktops

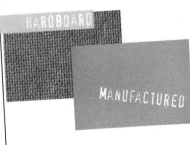

HARDBOARD

MANUFACTURED

WHY IT'S USED
Flat
Lightweight
Low cost

USES
Indoor furniture
Back of wardrobes
Bottom of drawers

Plastic

You should know about two main 'families' of plastics:

- **thermoplastics** – can be repeatedly heated and reformed
- **thermoset** plastics – can only be heated and formed once.

Within each family there are thousands of different types of plastic. Fortunately, you do not need to know about all of them! On the opposite page you will see some common plastics, their basic properties and uses.

Properties of Plastics

Although each type of plastic has its own unique properties, it is useful to understand some generic characteristics. A designer may choose to use a plastic over other materials because they:

- are lightweight
- show resistance to corrosion
- can be reheated and reshaped (thermoplastics)
- are translucent, transparent or opaque
- can insulate heat and/or electricity
- are easily formed
- are recyclable (thermoplastics).

Advantages of Plastics

Plastics may be obtained in powder, fibres, granules, liquid, film, clear or coloured sheet, rod and tube. They have various advantages over wood or metal:

- Coloured pigments can be added.
- They are naturally water resistant – no treatment is needed.
- Thermoplastics form easily and can be shaped.
- Suitable for mass production (high speed/high volume production).
- Insulator properties are useful if heat or electricity is present.

Quick Test

1. **State four** thermoplastics and their properties.

2. **Explain** why designers may choose to use plastics over traditional materials such as wood and metal.

ACRYLIC

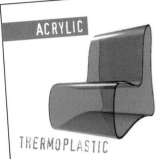

THERMOPLASTIC

WHY IT'S USED
Strong
Shatter resistant
Transparent

USES
Subsitute for glass
Car headlights
Shower doors

POLYPROPLYLENE

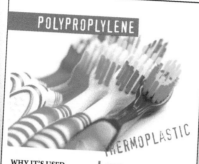

THERMOPLASTIC

WHY IT'S USED
Good chemical resistance
Resists fatigue
Light

USES
Rope
Toothbrushes
Medical equipment
Seats

PVC

THERMOPLASTIC

WHY IT'S USED
Durable
Rigid
Weather proof
Electrical insulator

USES
Window frames
Pipes and gutters
Cabling

HDPE

THERMOPLASTIC

WHY IT'S USED
Good chemical resistance
Durable
Tough

USES
Detergent bottles
Sterilised containers
Wheelie bins

ABS   THERMOPLASTIC

WHY IT'S USED
Scratch resistant
Light
Durable

USES
Toys
Crash helmets
Phones
3D printing

POLYSTYRENE

THERMOPLASTIC

WHY IT'S USED
Stiff
Light
Buoyant

USES
Food packaging
Toilet seats
Vending machine cups

UREA FORMALDEHYDE

THERMOSET

WHY IT'S USED
Electrical insulator

USES
Plugs and sockets
Electrical components

Fixing and Joining Techniques

By now, you will probably have experience in joining materials together. In industry, manufacturers also join materials together. There are two types of join:

- permanent
- non-permanent.

The choice of joining method depends on a number of considerations, including:

- Are the materials the same or different?
- Does the join need to be permanent or non-permanent?
- How strong does the join need to be?
- Should the join be visible or hidden?

Task 1

Carry out online research to find out different methods for joining the following materials together:

- wood and plastic
- wood and metal
- metal and plastic
- flat-pack furniture.

Remember, there are permanent and non-permanent joining methods. Try recording your findings on a 'mind map' to help revise the methods. Answers on page 144.

Task 2

Here are three different wood joints. Using the internet, find out:

- what these wood joints are called and their uses
- **four** other wood joints and their uses.
  Answers on page 144.

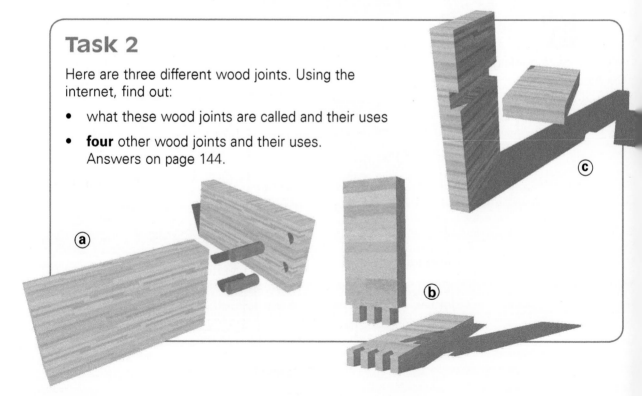

Manufacturing Processes

There is a vast range of commercial manufacturing processes for **forming**, **molding**, **cutting** and **casting** materials. The following table describes six manufacturing processes that you should have knowledge of for your exam. Bear in mind that you do not need to learn how each process works. You should, instead, focus on understanding:

- the materials used in each process
- the types of parts or products each process can produce
- how to identify the way in which a product has been manufactured.

Manufacturing process	Description	Example products	Weblink
Injection moulding	Creates complex, accurate parts from plastics such as ABS and Polypropylene.	syringes, toothbrushes, bottle lids, disposable razors	www.bpf.co.uk/ Plastipedia/Processes/ Injection_Moulding.aspx
Vacuum forming	Creates parts by stretching heated sheet plastic over a **pattern**. Plastics include ABS, Polypropylene and PVC.	yoghurt pots, chocolate tray packaging and sandwich boxes	www.bpf.co.uk/ Plastipedia/Processes/ Vacuum_Forming.aspx
Die casting	Creates complex parts where accuracy is important. Good surface detail can be achieved. Materials include aluminium and copper.	alloy wheels and car engine parts	
Rotational moulding	Creates hollow products from plastics such as Polypropylene and HDPE (High Density Polyethylene).	canoes and kayaks, traffic cones, rainwater tanks, children's playhouses	www.bpf.co.uk/ Plastipedia/Processes/ Rotational_Moulding.aspx
Extrusion	Creates parts with a uniform cross section with plastics such as PVC and HDPE. Metals such as aluminium, steel and copper can also be extruded.	drainage tubing, PVC window framing and gutters	www.bpf.co.uk/ Plastipedia/Processes/ Extrusion.aspx
Turning	Creates cylindrical parts of varying diameters. Can be manually turned or by CNC. Surface finish can be high.	precision engineered parts and cylindrical car engine parts	

Impact on Society

Design has changed society significantly over recent decades and will continue to shape consumer habits and lifestyles into the future. You should have an awareness of the impact of design technologies on society and the environment.

Question Paper Tip

You may be asked about the impact of technology on society. Try to learn some examples you could describe in the exam.

What Was Life Like?

In the 1940s life was very different to today. Women stayed at home to cook and look after the children (often more than three children); men worked and provided financially for their families. Money was spent on necessities, and products for the home were bought to last. Rationing was still in place due to the Second World War, meaning food and other products were all in short supply. Choice was not an option – people took what they could get. Items were bought mainly from local shops. There were fewer than 15,000 television sets in the whole of the UK!

Rise of Consumerism

Consumerism is an ever-increasing consumption of goods. Peer pressure and increased exposure to the media has created a 'want' society. People buy things to make them feel good or to be accepted, admired or envied by others. Christmas is a perfect example of an occasion that is dominated by consumerism. Santa Claus has been used by Coca Cola in advertising campaigns since 1931. Christmas has become a time for excessive spending and exchanging of gifts, with shops displaying seasonal products from October to cash in on what was once a more traditional and religious celebration.

PROS

As consumers buy increasing numbers of products, there is a growing benefit to the economy; trade creates jobs in manufacturing, retail and design.

CONS

Consumerism is the belief that everything can be bought and sold, an unhealthy view to hold and a habit that many families cannot afford to sustain. We often replace products for fashion reasons or due to enhanced technology – because we want to, not because we need to. This irresponsible consumerism is the root cause of many environmental and sustainability issues, including pollution and deforestation.

Affordable and Accessible Products

Advances in technology have continued to have an impact on the manufacturing industry.

Mass Production

Products can now be produced in high volumes and sold at a low cost (this is known as economy of scale). **Globalisation** has further enabled low-cost manufacture. Many high-street stores stock mass produced, low-cost fashion items. Consumers are able to stay in fashion, affording to buy and replace products more often.

Finance

Options such as credit cards, store cards, catalogues and hire purchase have made many products accessible to consumers on low incomes. Unfortunately, our growing need to 'keep up appearances' has left millions of consumers in debt with unmanageable repayments.

Media – growth in communications via satellite, TV, Internet and radio has opened up a global market, increasing consumers' awareness of and access to products. Similarly, access to the Internet and TV has also enhanced the consumers' ability to shop from the comfort of their home, making products more accessible.

Competition – consumers are now experienced shoppers and know what they want. This means that the product must meet the needs of the consumer in all its aspects. Discounts, sales and offers from competitor companies ensure that products are affordable, as businesses strive to keep their market share. Family cars and improvements in transportation have resulted in the development of large retail parks. As products become more accessible, the competition between companies also increases.

Planned obsolescence – companies design affordable products with a limited lifespan.

Impact on the environment – accessible and affordable products have obvious advantages to society and the economy; however, the irresponsible rise in consumerism has had negative effects on the environment and our sustainability. Designers must make responsible choices, from the sourcing of materials to designing products that are easily maintained or designed for disassembly.

Quick Test

1. **State three** lifestyle changes that have occurred since the war.

2. **List** the advantages and disadvantages of consumerism.

3. **List five** reasons products that have become more affordable or accessible

Impact of Technologies on Manufacturing

You need to have an understanding of the impact of technologies on manufacturing for both the Area of Study and for questions in the written exam. Manufacturing has changed a lot over recent decades and will continue to change as technologies emerge and develop in the future. Factories today bear little resemblance to those in the past. With such change come advantages and disadvantages.

Reduction in Workforce

Machines, CAD/CAM and **automation** in factories have allowed us to manufacture higher volumes with greater accuracy in less time, compared to using traditional manual methods. The downside of this is a reduction in workforce, as fewer people are needed to support the manufacturing process. Better communication technologies have opened up a global market and companies can locate wherever labour is cheapest. These changes have resulted in job losses in the UK as people are replaced by machines and cheaper labour elsewhere in the world.

Skilled Workforce

As automated systems like CAD/CAM have become more popular, the skills required by the workforce have changed greatly:

- greater demand for CAD operators
- greater demand for staff to supervise or maintain machines
- reduced demand for trade craftspeople.

Cost of Equipment

Manufacturing technologies come at a price. Companies have to make significant investment before they can make a profit. Investments include:

- purchasing and installing new machines
- maintaining highly technical equipment
- training staff in how to use the machines.

Impact on Environment

Manufacturing has negative effects on our environment. To manufacture products, companies use up the Earth's resources for energy and materials, generate waste, create pollution through manufacturing, transportation and disregarded products and packaging. This has impacted the environment and societies across the globe, and we are now dealing with the impact of global warming – a rise in average global temperature, leading to natural disasters and decreasing biodiversity as many species' habitats change or disappear.

Measures to Support Sustainability

Non-toxic substitutes
Optimise raw material use
Water use and waste water reduction
Air emission reduction
Solid and hazardous waste reduction
Transport package optimisation
Energy efficiency

Sustainability, that is using the Earth's resources responsibly and at a rate whereby the Earth will be able to provide for future generations, is everyone's responsibility. Designers and manufactures must aim to reduce the environmental impact of the products they design and manufacture to protect our environment and limit the use of resources. This is known as **clean manufacturing** and it includes sourcing materials sustainably and/or locally, and minimising transportation, energy consumption and waste from obsolescent or difficult-to-repair products. The NO WASTE (see the box, above right) acronym will help you remember the key points of a clean manufacturing system.

The 6 Rs are:
REDUCE
RETHINK
REFUSE
RECYCLE
REUSE
REPAIR

You may already be aware of the 6Rs (see box on right). Having an understanding of these will help you answer exam questions relating to the environment or sustainability.

Task – 6Rs

Create a 'mind map' exploring how the 6Rs apply to Designers, when making decisions during the design process; Manufacturers, when producing and distributing the products; Consumers, when using and disposing of products. Answers on page 144.

Online Resources

The Ellen McArthur Foundation has some excellent resources on the **circular economy**. The idea of a circular economy is that nothing goes to waste. Companies are encouraged to find ways to mimic the sustainable systems in nature, whereby all waste products get fed back into the cycle to support continual growth and maintain the planet's resources. There is also a focus on companies renting out products and offering 'buy-back' schemes to cut down on landfill from obsolete or unwanted products.

Link to Ellen McArthur Foundation
www.ellenmacarthur foundation.org/schools

Quick Test

1. How can clever use of packaging reduce the impact of transportation?

2. How has technology impacted the world of work?

3. What is the circular economy?

Practice Questions: Workshop

1. The small table below was produced in the workshop by a candidate.

a) i) **Name** a softwood that would be suitable for manufacture of the table legs. (1)

   ii) Give reasons for your answer. (2)

b) **Name** a dark coloured hardwood that could have been used instead of staining the pine tabletop. (1)

c) A corner-halving joint was used to join the legs and base (A–B), as shown in the photograph. **State** the name of an alternative joint that could have been used. (1)

2. A candidate manufactured the clock shown below using wood.

a) i) **Name** a suitable **manufactured** board that could be used to make the clock. (1)

   ii) **Give** a reason why the manufactured board you have chosen is suitable. (1)

b) **Describe two** methods that could have been used to manufacture the circular shapes (numbered parts) from wood in the workshop. (4)

c) **Describe** how the candidate could have removed the material to create the hole shape in the clock. (4)

3. The bedside unit below has been manufactured in the workshop.

Front bar

a) The candidate used blind holes in the centre of the drawer fronts to add detail. **Describe** how these holes would:

i) be marked out in the workshop (2)

ii) be manufactured/cut out in the workshop. (2)

b) The candidate has used a mortise and tenon joint to assemble the front bar into the sides. **Name** an alternative joint that could be used. (1)

A low-cost manufactured board has been used for the base of the drawers and the back of the unit.

c) **State** the name of the manufactured board that would have been used, giving a reason for your answer. (2)

4. The rocker below was manufactured in response to a design brief. The blue and yellow bars were manufactured from metal.

a) i) **State** the name of a suitable metal to manufacture the blue and yellow bars, giving a reason for your choice. (2)

ii) The bars have been dip-coated in plastic. State two advantages of dip-coating metal. (2)

b) The two rockers (green parts) need to be identical. **Describe two** methods the candidate could have used when manufacturing the rockers to ensure this fact. (4)

Practice Questions: Commercial

1. Standard NHS crutches are shown below.

 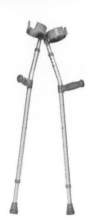

   a) **Explain** why function is more important than aesthetics in the design of a medical-aid product like crutches. (2)

   b) **State** the name of a metal and a plastic that would be suitable for manufacture of the crutches. **Give** reasons for your answers. (2)

   c) The manufacturer has used standard components and forms of supply. What are the advantages to the manufacturer of taking this approach? (4)

   d) **Describe** how the designer could evaluate the ergonomics of the crutches. (3)

2. A typical bagless vacuum cleaner is shown below.

   a) **State three** steps a designer could take to ensure products like this are designed with sustainability in mind. (3)

   b) The Dyson vacuum cleaner is a good example of a product that was designed using technology transfer. **Explain** what is meant by technology transfer and say how it can be used to help generate ideas for new products. (3)

c) When designing the vacuum cleaner, the designer made good use of modelling to refine the idea. **Explain** how the following models could have been useful during the design process:

  i) CAD (2)

  ii) parts model (2)

  iii) prototype. (2)

3. Products, like the concept car shown here, do not always make it into commercial production as they are considered too radical.

a) **Describe two** strategies that could be carried out by the marketing team to launch a new or unique product. (4)

Branding can have a great impact on consumers' perception of products and can affect their decision to invest in a product.

b) **Explain** the importance of branding and say how it influences consumers. (4)

4. A flexible camera tripod is shown.

Before a specification could be drawn up for the product, the designer carried out some initial research.

a) **State three** design factors that the designer could have researched before writing a specification for a camera tripod. (1)

b) **Explain** why these three factors are important. (3)

5. An upmarket pram is shown below.

a) **Describe** how ergonomics influenced the design of the pram. (4)

b) **Name** an appropriate method that market researchers could use to gather information on the following aspects of the pram:

i) function

ii) aesthetics

iii) performance. (3)

c) **Describe** how **one** of these methods could be used by the researcher. (2)

d) **Describe** an activity the researcher could use to carry out market research on the aesthetics of the pram. (2)

6. Designers and manufacturers have a responsibility to ensure they are designing and manufacturing sustainable products.

a) **State two** ways in which designers can influence consumers to become more responsible. (2)

b) **Explain two** strategies a manufacturer could apply to ensure sustainable manufacture. (4)

7. Some commercially manufactured parts and products are shown below.

i)

ii)

iii)

a) From the list below, **state** the manufacturing process that has been used to make each of the products.

Rotational moulding	**Die casting**
Injection moulding	**Vacuum forming**
Extrusion	

8. The coffee makers below have been manufactured using stainless steel.

a) **Give two** reasons why stainless steel is a suitable material for these products. (2)

b) **Compare** the aesthetics of these products. (6)

Answers

Workshop

1. **a)** **i)** Pine

    ii) Pine is light coloured, has a knotty appearance, is easy to work with and is easily available.

    b) Mahogany, walnut

    c) Dowel joint, (mortise and tenon).

2. **a)** **i)** Plywood, MDF

    ii) Plywood has a wood grain/nice aesthetic, is a strong and stable board. MDF has a smooth surface and edges, suitable for painting. A stable and inexpensive board.

    b) The candidate could turn the discs on the lathe. These would be created using one blank that would be turned down to the correct diameter, then they would use a parting tool. They could also make them by drawing circles onto sheet materials, cutting them out with a coping or scroll saw before sending them to shape and size.

    c) First the candidate could drill some holes inside the area they want to remove. If they hold the wood in a vice they can then take a coping saw, insert the blade into the hole and then reattach to the frame. They will then be able to saw the material out. When finished sawing, they can then move the blade and the coping saw before filing and sanding to shape.

3. **a)** **i)** Using a steel rule, the candidate could draw diagonal lines between the corners. This would give the centre of the drawer front. Alternatively, the candidate could use a steel rule to measure half the length of the drawer front. Using a try-square, they could then draw a vertical line from the halfway point. Finally, they would measure and mark the halfway point on the vertical line.

    ii) The hole could be produced using a forstner or flat-bottomed bit. The candidate would need to set the depth stop on the pillar drill to ensure they don't drill all the way through the wood.

    b) A dowel joint

    c) Hardboard because it is cheap, lightweight, normally used for drawers and furniture where it cannot be seen and usually has a decorative surface on one side.

4. **a)** **i)** Mild steel - cheap, strong (Aluminium – good strength to weight ratio, lightweight, widely available, easy to work with)

    ii) Improves aesthetics by adding colour, plastic is less cold to touch (better comfort for the handles), offers protection against corrosion.

    b) Method 1: They could make a template in the shape of the rocker out of paper, card or thin sheet material. This could be used to trace around the shape onto the material, making each one identical.

    Method 2: The candidate could use masking tape or cramps to keep both pieces of work together so they can be cut and sanded simultaneously.

Commercial

1. **a)** The crutches are designed to help support people with an injury or disability, whilst walking or standing. This is the primary function. The aesthetics of the product are not essential as the products are often loaned to patients for short-term use while they recover.

   b) Aluminium – good strength to weight ratio, low maintenance, silver colour (stainless steel); ABS/Polypropylene – durable, suitable for mass production

   c) The manufacturer will save money using standard forms of supply as they are cheaper than getting materials supplied in custom sizes. Standard components also save the manufacturer money as they are cheaper than producing specialist parts, as they just buy them in. They don't need machines or staff to make them. The manufacturer can also save time using standard components as they just buy them in.

   d) Best method is a user trial: they could ask a range of different people to carry out a task like walking around a room, up and down stairs and getting in and out of a chair. They could observe the users using and adjusting the crutches then ask them about their experience, e.g. were they adjustable to a comfortable height, was there enough grip, were they comfortable to hold and around the arm? By observing and talking to the people carrying out the user trial, the designer will identify strengths and weaknesses of the product.

2. **a)** Use materials from a sustainable or local source, design the product for disassembly, use recyclable, biodegradable, bio materials, clear instructions for recycling and material ID, use standard components for easy maintenance (increase lifespan)

   b) Technology transfer is when an existing technology is taken from one product and applied in a new context to solve a different problem. Designers can use this to generate new ideas by thinking about their problem and recognising existing technology in different products or by taking an existing technology and finding a new and alternative use for it.

   c) i) CAD models could have been used during the development to visualise different colour options quickly. CAD models also allow the designer to communicate a lifelike image of the product to the client. This is particularly useful with products that are very small as they can zoom in on them (or out on very large products). The model could be used to generate 3D printed parts or to show or test how components would fit together. CAD enables the designer to run animations and simulations to show the client aesthetic options, how the product assembles and functions, without having to spend any money on manufacturing materials.

   ii) Parts models could have been used to check the flexibility of the head, or to figure out the best way to join several parts. The designer uses these when focusing on a specific problem, as there is no need to waste time making the other bits. They could test the head to check if it loses contact with the floor when used at an angle, like under the sofa. This would allow them to identify problems and make changes.

   iii) A prototype would be used to demonstrate how the final product works and looks. It can be used to evaluate the final design and make necessary changes before it goes into production because it will be fully functional and made from the correct materials.

3. a) They could carry out promotional activities to gain interest in the product. This could include special offers, free trials or other discounts/ privileges to encourage consumers to buy into the product. They could also use the media to build hype about their product. Many companies do this by keeping their products and release dates secret, leaking only some information. They then have an official launch event. At this point, there is already a lot of public and media interest and consumer demand for the product.

b) Consumers can buy branded products for social status or acceptance. Wearing or owning brands e.g. clothing, bags, cars etc. can make people feel good. Branding also communicates perception of quality, durability and value for money. If a range of identical products is being sold at similar prices, e.g. toasters, consumers are likely to buy the brand with which they have had a positive experience, rather than take the risk with an unknown brand. Branding also offers the consumer some reassurance that the product will be safe, regulated and of acceptable quality.

4. a) Any 3 from function, ergonomics, aesthetics, performance, cost/environment, market.

b) Function could have been researched to understand how the tripod might work. They need to ensure it will hold the camera securely and not let it fall. What size or range of camera will it hold, how will it stand or hang, is it fit for purpose?

Ergonomics would have been researched to ensure it was comfortable and easy to use. The designer would need to consider the strength of the users to operate the product and ensure it was easy to attach and remove the cameras.

Aesthetics could have been researched to ensure the product is gender-neutral. The aesthetics should also be easy to maintain as it will be used outdoors. The designer may have researched colours, as well as ways to make it look easy to use.

Performance could have been researched to look at durability of materials and components, finishes for ease of maintenance and options to improve ease of use.

Cost/Environment could have been researched because sustainable design is important. This means finding sustainable, renewable, recycled materials and ways to design for disassembly. The overall cost is important as volume and materials will affect it and the selling price needs to be greater to make a profile.

Market could have been researched as the designer may check out the competition, research what they like, where they shop, how much they would be prepared to spend, to make sure the product is suitable. The designer would need to understand the market to know where and how they would intend to use the product.

5. a) The height of the handle would have been influenced by ergonomics to make sure it is a comfortable height for users to push, ranging from small to tall individuals. The width of the seating area would have been considered to make sure that it was a comfortable size for toddlers to fit in. The position and height of the hood would have been considered because the pusher needs to see over it and it should have clearance space over the child's head. The straps on the pram would be influenced by ergonomics as they need to adjust to different body sizes to secure different sized children as they grow.

Any other ergonomic aspect, including: weight of pram to push, carry, transport, size of pram so easy to lift and move, grip size of handle.

b) i) Questionnaire, survey or user trip/trial

ii) Questionnaire or survey

iii) Performance – test, questionnaire, survey

c) The designer could use a questionnaire to find out about the function and performance of the pram. They could target new parents and ask them which features they would like the pram to have. They could also ask experienced users about the performance of their prams to find out how long it should last or how the user would maintain it. The answers would then be collated.

d) They could do a survey asking the potential market what they thought of the colour and style of the pram. They could also ask if it looks easy to clean, if they would use it for a boy or a girl and if they would buy it. All of these would give ideas on what the surveyed people thought about the aesthetics.

6. a) Make recycling options clear on the product, design for disassembly, use recyclable or biodegradable materials, offer updates to avoid premature disposal of functional products, use standard components to ensure it's easy to maintain to increase durability.

b) Any which explain: reducing the waste, using materials that produce minimum toxins and pollution, sourcing from local suppliers to reduce transportation pollution, minimising packaging.

7. a) i) Injection moulded

ii) Vacuum formed

iii) Rotational moulded

8. a) It resists corrosion so is suitable to be in and out of water, it has good strength and durability, Silvery colour is attractive, smooth surface is easy to wipe clean.

b) The smooth rounded shape of the second product makes it look simple to use and easy to clean. The flat faces and angles on the first product add visual interest and make the product look more expensive. Both products are made from silver-coloured metals with contrasting black plastic handles. The handle on the second has a smooth rounded form to compliment the main body. The flat wide bases on both models make the product look stable and safe to use. The smooth texture of both makes them look clean and easy to maintain.

Glossary

A

ABS – A thermoplastic commonly used for its impact resistance and strength. Also known as Acrylonitrile Butadiene Styrene.

Accessible – The degree to which a product is available to the widest range of users. A product could be made accessible in many ways, including size and price. Acrylic – A thermoplastic that is often used as an alternative to glass.

Aesthetics – A philosophy of the nature and the appreciation of beauty. Aesthetics involves our senses and our response. Do we like the look, feel, etc? Aesthetics is an important design factor to consider.

Aluminium – A non-ferrous silver-coloured metal used for its good strength to weight ratio.

Analogy – An analogy is a comparison between one thing and another. In idea generation, an analogy is used to compare your problem to a similar situation in a different field, to stimulate new solutions to the problem.

Analysis – The process of breaking down a complex problem or object into smaller, more manageable parts to explore and gain a better understanding of it.

Annotations – Written notes added to further explain or elaborate on an idea.

Ash – A hardwood commonly used for sports equipment and steam-bent products due to the high flexibility of the timber.

Assembly – The process of joining two or more parts together.

B

Beech – A hardwood known for its strength and food-safe properties.

Brainstorming – A group-based idea-generation technique whereby thoughts and ideas are quickly shared and recorded. No ideas are discarded.

Brief – A statement written at the beginning of the design process by the client or designer, specifying the details for a design task.

C

CAD – Computer Aided Design

Cedar – A softwood that is resistant to weather and is lightweight.

Circular Economy – An approach in industry used to keep resources and materials in circulation, rather than disposing of them. A circular economy builds sustainability.

Client – A person or organisation paying for the services of a professional.

Commercial Manufacture – The industrial production of products, made with the intention of being sold. Specialist machines and processes are required to produce a large volume of products quickly.

Communicate – To express/share thoughts, ideas, decisions and reasons. Photographs, models, written comments and sketches are some of the ways in which you could communicate your design work.

Consumer – The person who buys or uses a product or service.

Consumerism – The behaviour of buying, using and disposing of goods.

Copper – A non-ferrous metal used for its conductive and ductile properties.

Cutting List – A written list of materials, parts and sizes needed to manufacture an item, usually produced as a table.

D

Deforestation – When significant areas of forests are cut down and the land used for other purposes.

Design Factors – Factors that influence the design of a product, such as aesthetics, ergonomics and function.

Design Process – The name given to the different stages of designing, from problem to solution.

Designer – A person in the design team who produces design work, often in response to a design brief.

Development – The process of exploring and refining an idea towards a design solution.

Dimension – A size indicated on a drawing.

Ductile – Material property: the ability of a material to be stretched without breaking.

E

Emery Cloth – Abrasive sheets with a fabric back, used to rub down metals. Similar to sandpaper, different grades of coarseness are available.

Ergonomics – The design factor that looks at how people interact with products and their environment.

Evaluation – The process of judging ideas against a criteria or specification.

Exploded – A pictorial drawing type where all the parts of an object are separated to show how they should be assembled.

Explore – To investigate aspects of your design, considering alternative solutions to identify the best one.

F

Function – The design factor covering everything related to what a product does. Can be divided into primary and secondary functions.

G

Gender – The term given to being male or female.

H

Hardwood – Wood that comes from a broadleaved tree. Hardwoods are used often for furniture and flooring.

HDPE (High Density Polyethylene) – A thermoplastic commonly used for its strength and stiffness.

I

Idea Generation – The process of creating ideas.

Injection Moulding – Commercial manufacturing process whereby molten plastic is 'injected' at high pressure into a mould. Used in high-volume production.

Iron – A hard, heavy and brittle ferrous metal.

K

Knock-Down Fittings (KDF) – A group of fixtures commonly used in flat-pack furniture to join parts together. Makes products quick and easy to assemble.

L

Lateral Thinking – Looking at a problem from a different perspective and producing unconventional ideas.

Lifestyle Board – A collection of images used to communicate the lifestyle of a target market. Information could include gender, age, family, employment, interests and style of the market.

M

Mahogany – A dark brown coloured hardwood.

Market – A group of people who are potentially users of a product or service.

Modelling – The process of producing a three-dimensional physical artefact that can demonstrate, for example, the size, proportions and materials of an idea.

Mood Board – A collection of images, materials or text that communicates the intended style or theme of a product.

Morphological Analysis – An idea-generation technique that allows ideas for products with many combinations of features to be produced.

O

Oak – A commonly used hardwood known for its hard and strong properties.

Obsolescence (Planned) – Condition of manufactured items becoming discarded or outdated in a set time, for reasons including new technology or the manufacturer discontinuing production of the product or its spare parts.

P

Performance – How well a product carries out its job and meets the expectations of the user in ease of use, operation, maintenance and value for money.

Pictorial – A type of drawing that appears to be three-dimensional on paper. Examples include isometric and two-point perspective.

Pine – A light yellow coloured softwood commonly used in furniture. Normally cheaper than hardwoods.

Polypropylene – A thermoplastic commonly used in domestic appliances, furniture and medical applications due to good chemical and heat resistance, good fatigue resistance and its integral hinge property.

Procedure – A step-by-step list detailing the process of manufacturing a prototype.

Properties (of Materials) – Attributes used to describe the characteristics of materials, e.g. ductile.

Prototype – A full-scale model of a developed design, made from final materials and used to test and evaluate against a specification.

PVC (Polyvinyl Chloride) – A thermoplastic commonly used for its toughness and durability.

Q

Questionnaire – A paper-based research method used to gather information from a target market, user or client.

R

Refine – Part of idea development; when the finer details of the design are established.

Research – The process of finding out a range of information relevant to a design task.

S

Sequence of Operations – A list of steps, devised pre-production and in a logical order, that are required to manufacture a product.

Softwood – Wood from a fast-growing conifer tree, often from sustainable forests and usually less expensive than hardwood.

Specification – A list of requirements for a design solution.

Stain – A coloured, protective finish that can be applied to wood. Usually applied with a brush or cloth.

Steel – A ferrous metal used for its hardness and tensile strength (available as mild, stainless and high-carbon steel).

Survey – A list of questions asked by market researchers to gather specific information about market, consumers, retailers, etc.

Sustainability – Using the Earth's natural resources in such a way that they are not depleted for future generations.

T

Technology Transfer – The process of transfering existing technologies into new products or applications.

Thermoplastics – A family of plastics that become soft and malleable when heated, and which set when cooled.

Thermoset Plastic – A family of plastics that can only be formed once through heating.

U

Urea Formaldehyde – A durable thermoset plastic commonly used for its good thermal and electrical resistance.

W

Walnut – A hardwood known for its dark colour and tight grain, traditionally used for carved furniture.

Wet and Dry – An abrasive coated paper used on plastics and wood. Similar to sand paper but much finer.

Working Drawing – A technical drawing of a product, with information such as dimensions, scale and orthographic views.

Answers to Revision Questions

Design Factors – Page 43

1. The primary function of the camping stove is to heat/cook food.

2. The camping stove would not be fit for purpose if it were used in a scenario other than for camping (e.g. if it was used at home as a family cooking stove).

3. Value for money – if it is affordable to buy, run and maintain.
Planned obsolescence – how long the product will last in comparison to how long the consumer would want to use it.
Ease of maintenance – how easy the product is to clean and maintain.
Durability – how long the components last, how resistant to damage or wear.

4. Using standard components, selecting materials that can be cleaned easily, designing casings that can be opened to allow internal parts to be replaced.

5. **a)** Advantages – A greater choice of products is available on the market. Enables them to stay up to date with new products and technology.
Disadvantages – They spend more money on upgrading products. They may feel under pressure to constantly upgrade their products, even if they still work perfectly well.

   b) Advantages – They receive more repeat business from consumers. Brand loyalty and awareness improves as they appear to be at the forefront of technology.
Disadvantages – They have continual research and development costs, repeated risks with new product launches.

6. The handle of the hairdryer has a curved wavy form. This allows the user to have a better and more comfortable grip. The nozzle changes form towards the front. This change pushes the air out through a smaller space, improving the function of the hairdryer. The rectangular shape of the buttons contrast with the curvy form of the hairdryer. This makes them stand out.

7. Contrast, harmony, colour, texture, proportion, size and materials.

8. Reach/shelf – To find the correct height and depth to ensure the majority of people can reach and use it.
Clearance/desk –To find the seat size and largest thigh height, to ensure that the majority of people can sit under a table with room to move their legs.
Grip/bike – To ensure the user's hands don't slip off the handles or the brakes.
Weight/elevator – To ensure that the lift can move the maximum recommended weight of people.
Strength and effort/medicine bottle – To ensure that children don't have the strength required to press and open medicine bottles.
Body dimensions/Xbox controller – Finger sizes, widths to ensure clearance between buttons and length to ensure they could reach the buttons.

9. The width of the human palm would help to determine the length of the purple handles. The distance between the palm and finger will determine the distance between the handles.

10. The designer could use straps that can be adjusted to fit different sizes of heads.

11. **a)** Consumer demand is what the consumers want and are asking for, e.g. more environmentally friendly products or phones with more memory, faster internet etc.

   b) Market segment is the term used when a market is broken down into different areas. For example, the market can be broken down into segments by age, gender, geographic location or income.

   c) Technology push is when new technology is brought to the market, e.g. the first mobile phone or the sony Walkman.

   d) Needs and wants are different things. Needs are essential things that people require to live a normal life. Wants are desirable things that people would like, but do not necessarily need.

   e) Branding is the name, logo or symbols used to identify a company.

Design Process – Page 56

1. Designer – To generate and create solutions.
Market researcher – Knows the market. Support: Informs designer of needs and wants of the market.
Accountant – Controls finances. Support: Monitors the budgets and profits.
Engineer – Specialises in areas of design and manufacture. Support: Advises on structural and technical aspects of the design.
Manufacturer – Makes the product. Support: Advises the best way to manufacture the product.
Ergonomist – Specialises in human data/engineering. Support: Helps the designer with ergonomic aspects of the design.

2. User trip, user trial, questionnaire, survey, focus group.

3. Media advertisement (TV, radio, magazines, press), offers – discounts, sampling, free trial etc.

4. Working drawings, technical, exploded detail, model showing assembly of components.

5. Designer can fully understand the needs of the client. Clear up any uncertainties.

6. Provides details that can inform the design and the specification.

7. Testing – An activity that only requires one result, e.g. checking to see if a switch works, or if a kettle boils water.
Questionnaires – Gather a range of opinions, e.g. information on consumer expectations or experiences.
Search engines – Compare products, data on sales, material qualities, market.
Measuring/recording – Length of cables, user opinions, range of opinions.
Using data – Data tables, including material properties and ergonomic factors.
User trip/trial – Gather information on the use of the product by using it or observing others

8. The specification will lack distinct detail. It will be generic or based on guess work if no research or poor research has been carried out. The same applies if no conclusions are drawn from the research or if the information is not included in the specification.

9. List of what the product MUST do. Used as a checklist for the designer to evaluate their designs.

10. Initial ideas: to evaluate the success so far and identify the ideas with most potential.
Development: to target areas for improvement.
Proposal: to evaluate the design proposal.

11. **a)** Technology transfer: take an existing technology or system and apply it in a new context to solve a design problem.

   b) Biomimicry: Look to the systems and structures in nature for inspiration. Recreate these to solve design problems

   c) SCAMPER: Use the attributes of each letter to develop and explore an idea.

   d) Six-hat thinking: Adopt the different mindsets to critically evaluate and assess an idea, to inform the development.

12. Advantages: Quick and easy to use.
Disadvantages: Limited to only generating irregular 2D shapes.

13. Advantage: Can identify the strengths to keep and the weaknesses to develop. Disadvantages: Can copy designs, meaning they are not original.

14. Initial ideas: To evaluate the success/potential of an idea.
Development: To evaluate the success/potential of an idea.
Proposal: To evaluate the success of the proposal or manufacturing plan.

15. The aesthetics or proportions of a design, the ergonomics or function of a design, the working parts or assembly of a design.

Planning for Manufacture – Page 87

1. Dimensions (sizes); scale; related orthographic views; pictorial and exploded views; technical detail.

2. Prior to the manufacture of the product.

3. The information on a working drawing can help ensure accuracy of the final model (particularly for measuring and marking out of material).

4. Exploded views: Show how parts fit together.
Detailed views: Parts are enlarged so they can be visualised more easily.

5. The length, breadth and thickness of each part; how many of each part you need to manufacture; the type of material used.

6. It will be unclear as to the tools required, how parts fit together and the order in which to carry out the manufacturing steps of the model/product/prototype. This could impact on the time it takes to manufacture and also the accuracy of the final product.

7. Sequence of operations, cutting list, working drawing.

Workshop Manufacture – Page 94

1. Pencil; steel rule; marking gauge; mortise gauge; try-square; marking knife; sliding bevel; mitre square.

2. Scriber; steel rule; engineers try-square; spring dividers; odd-leg callipers; centre punch.

3. Check against the working drawing; measure again.

4. Bevel-edged chisel; mortise chisel; tin snips; guillotine; scissors; pillar drill; mortise machine; planes; files.

5. Shaping is performed by removing material using cutting tools, glass/sand paper, etc. Forming is when material is shaped or folded around a mould. Shaping is 2D; forming is 3D.

6. Vacuum former; strip heater, plastic oven.

7. Wood: Apply sand/glass paper by gradually moving up the different grades of paper to wet and dry. Metal: Filing, emery cloth, polish. Plastic: Cross file, draw file, wet and dry, polish.

8. Wood: Stain, die, paint, varnish, lacquer, oil, wax. Metal: Paint, dip coating.

9. Ensures accuracy, especially for multiple parts that are the same; easier to hold.

10. Wood joints; screws; knock-down fittings; adhesives; nails; hinges; rivets; nuts and bolts.

11. Safety goggles; ear protectors; visor; dust mask; apron; leather apron; heat-resistant gloves.

12. Dowel joint, corner lap.

13. Steel rule, try-square, marking gauge, marking knife.

14. Tenon saw, chisel.

15. Try-square to check corners are 90 degrees, measure/check the diagonals.

16. Wood dye, wood stain (paint is not an acceptable answer as it would hide the wood grain).

17. Oven, strip heater.

18. Always support, e.g. place scrap wood underneath and drill slowly, or use a step drill, or apply masking tape to the surface.

Commercial Manufacture – Page 106

1. Softwoods: Pine, spruce, cedar.
   Hardwoods: Beech, mahogany, oak, elm, ash, teak.

2. Softwoods normally grow quicker than hardwoods, meaning trees that are cut down can be replanted and re-grown within a shorter period of time.

3. MDF, hardboard, chipboard, plywood.

4. Available in large, flat sheets; available in a range of uniform thicknesses; low cost compared to most natural timber; boards are very stable.

5. Beech.

6. Red cedar is weather-durable (due to the natural oils within the material).

7. Ferrous.

8. Aluminium is lightweight, safe to drink from, is suitable for forming and is odourless.

9. Extrusion.

10. Resistant to rust; durable; silver colour provides good aesthetic qualities.

11. Thermoplastics can be reshaped and reformed through heating. Thermoplastics can be recycled. Thermoset plastics, once formed and cooled, become permanently solid and cannot be reformed or reshaped.

12. Thermoplastics: Acrylic, polypropylene, polystyrene, polycarbonate, polythene, HDPE, LDPE.
    Thermoset: Epoxy resin, urea formaldehyde, melamine formaldehyde.

13. Injection moulding; extrusion; rotational moulding.

14. Strong; easy to form; food safe.

15. Complex parts can be created; good surface finish can be achieved (without the need for further finishing).

Answers to Quick Tests

Page 45

1. Adds multi-functionality to a product; makes a product more appealing to the consumer.
2. To be fit for purpose, a product must work and be suitable for its intended use.

Page 47

1. Ease of use, ease of maintenance, durability and planned obsolescence.
2. Consumers can throw away perfectly functional products that may end up in landfill.

Page 53

1. Shape, form, proportion, colour, texture, materials, style and fashion, contrast and harmony.
2. A consumer's first impression can influence their choice to buy a product. A designer needs to know what style, colour etc. will appeal to the target market.

Page 55

1. A target market is a specific group of people, like dog walkers or swimmers, and a niche market is a combination of the product and target market.
2. Branding can influence how likely a consumer is to buy a product. A brand can suggest quality or provide buyers with a sense of social status.

Page 60

1. Carry out situation analysis, identify needs and wants, and carry out a product evaluation.
2. Comparing similar products, testing a product, observing people using a product (and recording their feedback).

Page 61

1. To give the designer a starting point; provides the designer with a clear statement of the problem or the situation; explain what is required for a design solution; clearly indicate who target market is.
2. Allows designer to explore possibilities for different aspects of the brief.
3. Ask a series of questions relating to the brief, such as: Where is the product going to be used? Who is the product aimed at? What is the function of the product?

Page 63

1. Asking experts, questionnaires or surveys, internet search, books/magazines/journals, observations, testing, data tables, measuring.
2. Research is necessary to allow you to add detail to the brief, by writing a product design specification. The specification requires facts from your research.
3. The majority of commercial products that fail do so because of lack of appropriate or meaningful research. It ensures a product responds to the design brief and is suitable for the target market.

Page 65

1. There is no value in writing about materials during research unless the information will be used and referred to when developing the product.
2. To inform the product development.
3. The designer is not designing for themselves. Some research requires factual information. An opinion of one person is not an appropriate method.

Page 67

1. It clarifies what the solution must do and allows the designer to evaluate their decisions and ideas to ensure they are on the right track to developing a good solution.
2. Using the results of your research and the design brief.

Page 71

1. Annotations, pictorial and 2D sketches, modelling.
2. To add more detail and/or clarity to a sketch.

Page 73

1. Idea generation, exploring or developing an idea, refining an idea, proposal. .
2. Generate ideas, visualise, test working or component parts, solve ergonomic, functional and aesthetic issues.

Page 74

1. Advantages – Files can be sent to a CNC machine for manufacturing; multiple changes can be made quickly and easy; actual materials can be applied digitally; uses no material; multiple copies can be saved and emailed anywhere; workshop environment and equipment is not needed. Disadvantages – High set up costs, staff training, can't touch, feel and interact with the CAD model, the way you can with a physical model.

2. Test model; scale model; 3D CAD model; prototype; mock-up.

Page 77

1. Helps you generate unusual ideas that you would not have thought of otherwise. Ideas are usually more creative.

2. Taking your pencil for a walk, morphological analysis, brainstorming, technology transfer, analogy/biomimicry, mood/lifestyle boards, SCAMPER, six-hat thinking.

Page 83

1. Sizes for all components, sizes of drills to be used, shape of components.

2. Check your solution meets all requirements from the specification; finalise and communicate all sizes to ensure manufacture can take place; make and record informed decisions of the materials you require; make and record informed decisions about the assembly methods.

3. It allows you to draw conclusions and informs your design decisions.

Page 85

1. There are no explanations of the decisions made. Star ratings do not communicate an understanding.

2. The specification.

3. Continually evaluating and reviewing their design decisions will keep the designer on track for developing a successful solution that appeals to the needs of the target market.

Page 89

1. Missing parts or dimensions will prevent the manufacturer from being able to manufacture the product.

2. Details of how the product will assemble; the main sizes of the assembly and component parts.

Page 91

1. A list of all of the parts, their sizes and materials that are needed to manufacture a product.

2. A sketch with all parts separate; using CAD software; creating a physical model showing all the parts separated.

Page 92

1. Prior to manufacture.

2. Order to carry out steps of manufacture; tools required; how parts fit together; type of finish required for each part.

Page 97

1. The overall accuracy of the manufactured model will be greater.

Page 101

1. Stain, die, wax, lacquer, varnish, paint, oil.

2. Use a coarse sandpaper (around P60) to remove pencil marks or major rough areas in the wood by sanding along the grain, not across it. Repeat this process with a higher-grade paper (around P120) to achieve a smooth finish. This process could be repeated again with an even higher grade paper before moving on to wet and dry paper. Using a sanding cork would be useful.

Page 103

1. Manufacturers buy the fittings in bulk quantities, reducing the costs of each unit part.

2. Costs are reduced. Easy to transport flat-pack furniture and easy to assemble.

Page 105

1. Companies have a legal requirement to provide a safe working environment for their employees. Unsafe environments lead to injuries and time off work.

2. Product testing (physical or CAD simulation); include safety labelling on packaging; quality assurance checks.

Page 112

1. Acrylic – Strong, transparent, chemical resistant. ABS – Impact resistant, strong, durable. Polypropylene – Shatter resistant, stiff, heat resistant. HDPE – Stiff, strong, tough, easy to form. PVC – Tough and durable, flexible, weather resistant, non-toxic.

2. They are naturally water resistant (no finish required); easy to clean; come in a range of colours; thermoplastics can be easily formed; good insulation properties; can be moulded.

Page 117

1. More women now work outside the home; choice and luxury items are now options for consumers; products and produce can be bought in supermarkets and online.

2. Advantages: There is a growing benefit to the economy, consumers now have more choice. Disadvantages: Can potentially be unaffordable to sustain; regular discarding of products has a negative effect on the environment.

3. Finance; media; competition; mass production; planned obsolescence.

Page 119

1. Lightweight packaging will reduce carbon emissions transportation. Where possible, reducing the size of packaging will mean that more products can be shipped at a time.

2. Machines, CAD/CAM and automation means there is a reduction in people required to work in factories (different skills are required now). Increased communication has opened up a global market.

3. The circular economy is a sustainability model where companies produce no or very little waste.

Answers to Tasks

Page 49: Ergonomics

Task 1

a) Users will require a certain amount of strength to press and hold the trigger/button; b) users need to reach with their index finger to pull the trigger/ button; c) the length of the handle will be determined by the width of the human hand; d) the battery will be heavy, therefore it is likely to cause fatigue.

Page 50: Ergonomics

Task 2

1. 95th percentile male hand breadth (98 mm) plus clearance.

2. Weight is important because, if the iron is heavy, the user's arm may become tired when using the iron for extended periods of time.

3. Modelling would let people pick up the product and see if their fingers could reach the buttons easily. The designer could get a range of users to try it to see if they can comfortably reach the buttons, as well as feeding back on the comfort of the form.

Task 3

Ergonomics has influenced the controller in lots of ways. The handles are shaped to allow the user to have a secure grip of the handset, while they switch between controls. The curved shape helps with comfort. The coloured buttons are positioned so that they are within easy reach of the user's thumb on their right hand. The user needs to be able to switch between them and the other controls quickly and easily. The spaces between the buttons help this as the user is less likely to press the wrong button by accident.

The joystick buttons have a recess on the top that makes it less likely for the user to lose grip during gameplay, as these are controlled by the the player's thumbs. The triggers are positioned at the back of the controller, meaning the designer would need to ensure that all of the users could reach the triggers when holding the handles.

Task 4

Adjustable seat height: body dimensions (leg length), reach (to reach pedals), strength and effort required to alter gear.

Water-bottle holder: arm length/reach, clearance from frame/holder and legs, strength and effort to remove and insert bottle.

Handlebars: body dimensions (grip size, hand width, arm length for position), reach from seated.

Brake levers: body dimensions (finger length, hand width), reach of fingers, fatigue of hands if too stiff to pull and hold, strength and effort required to engage break.

Page 53: Aesthetics

Task 1

The left-hand suitcase uses a silver colour for the body, which makes it look sophisticated. The contrasting black colour has been used to highlight functional parts, like the handle, wheel and locks. The bright pink colour used in the right-hand suitcase stands out, making the product appear more fun and vibrant than the silver suitcase.

The silver suitcase has a subtle embossed pattern of tall rectangles on the front face of the case. This makes the product more visually interesting and seem expensive. The pink suitcase is covered in a repeat pattern of hearts, which may make this case more distinctive and characterful.

Task 2

Guitar 1 – The black and dark red colour scheme makes the guitar look powerful and aggressive; the harmonising shapes of the body and pick guard look elegant. Guitar 2 – The bright orange and red colours make the guitar look warm; the contrasting white pick guard against the body stands out and looks exciting.

Page 55: Branding

Consumers can be loyal to a successful or reliable brand.

Consumers can be loyal to a successful or reliable brand for many reasons. Consumers may even limit their own choices by being loyal to a brand. For example, in the mobile phone market, consumers may choose to go back to the same brand (for example Apple), instead of exploring other products on the market. Consumers may be loyal to Apple because they get used to, like and desire the range of services, the user interface of the software and software updates, and the robustness of the products, as well as their reliable performance and crisp aesthetic.

Consumers can be equally loyal to lower-end brands. Primark is successful for selling affordable fashion items; Kia is growing in popularity for low-cost cars with extensive warranty. All of these are successful brands, as they cater for the needs of their respective markets. In today's competitive market, a company or brand must continue to adapt and respond to consumer demands and needs to maintain customer loyalty.

Branding can affect consumers' perception of the quality of a product.

If there were no brands, it would be very difficult for consumers to know what they were buying. We naturally base our decisions on past experience. If we bought a new laptop that didn't work and, say, the company chose not to replace or fix the product, would we want to buy from them again? The answer is no, but how would we know if nothing was branded? It is common for consumers to assume that unbranded products are inferior or of poor quality, although this is not always the case.

Branding is important because it gives us information about and confidence in the product we are buying. For example, if we buy a Sony laptop, the brand (and price) would suggest we are investing in a quality product. The consumer can check reviews for the company and product, and may also have past experience of Sony's other products. Branding allows consumers to connect products with aspects they feel will benefit them. As a result, consumers will buy into trusted brands, expecting them to deliver what they promise.

Branding can satisfy consumers' social and emotional needs.

Branding can provide consumers with a sense of wellbeing, satisfying emotional needs. Successful brands know how to connect with their consumers. As well as providing reliable products and services, successful brands gain loyalty by fulfilling our higher level needs, including for self-esteem and social interaction.

These underlying emotional needs can motivate consumers to buy. Brands can exploit these needs in a variety of ways. For example, a brand like Starbucks may appeal because the consumer believes in and wants to be seen supporting fair-trade companies. This makes consumers feel good and supports their social status.

Consider the BMW brand. A consumer may gain self-esteem because it gives them a sense of achievement to drive a quality, reliable and prestige product. Some consumers are motivated to buy things to 'show off' to others, rather than to satisfy their own requirements.

Branding is a powerful tool. It impacts how we feel and think about products and, sometimes, about the people who buy them.

Page 63: Research

1. The type of light switches, what teenagers will pay (stated in brief) and different wood metals and plastics.

2. What the teenager will use the light for – function
   What size is a 28 watt bulb – function
   What is the style of the Creative Currie range – aesthetics
   What aesthetic features will appeal to the teenage market – market/aesthetics

3 and 4. What the teenager will use the light for – questionnaire/ survey – because you can ask users for opinion. Size of a 28 watt bulb – measure bulbs or find online – because it will give accurate data. Style of the Creative Currie range – online research or visit – because you can get images of style. Aesthetic features that will appeal to the teenage market – questionnaire/survey – because you can ask the market for their opinion.

Page 71: Communication

1. Pictorial sketches (oblique, isometric) orthographic, assembly/ exploded detail, modelling and annotations.

2. Yes, because they communicate enough to clearly understand each change.

3. Decisions have been highlighted and are clear.

4. There are lots of reasons given. Ticks and crosses help communicate some decisions.

5. Annotation is sufficient. No improvements needed as it is clear and relevant.

6. The page has been split into three to separate different parts of the development. Arrows are used, providing a path to move through the development.

Page 78: Idea Generation

A suggested solution to this task can be viewed by following the weblink below.

> **Link to children's storage solution worked example**
> https://collins.co.uk/pages/scottish-curriculum-free-resources

Page 88: Working Drawings

There are three missing dimensions for each part: position of the centre of both semi-circles; position of both cuts; radius of both semi-circles; vertical position of the chamfers (diagonals).

Page 93: Evaluate the Plan for Manufacture

1. The dimensions are missing for the legs; no dimensions for the shape of side part; no dimensions for the position and length for both slots: no dimensions for the slots on the seat; no dimensions for the slots on the back; the width of the back is missing.

2. The width of the seat should be 50 mm, not 20 mm; the length and width of the back are missing.

3. There are no dimensions or instructions for marking out the side part; the dimensions for the back are missing from the sequence of operations; the bottom slot and top slot are not explained clearly.

4. Scissors are missing from the tools list; no information on what to use to colour in the side parts red; no information on what to use to colour the seat and back green.

Page 96: Measuring and Marking Out (Wood)

Measuring distances and using as a straight edge for marking out – Steel rule. Marking a line parallel to the edge of the wood – Marking gauge. Marking a line at right angles to the wood – Try-square. Mark two parallel lines to the edge of the wood for a mortise – Mortise gauge.

Page 97: Measuring and Marking Out (Metal)

Marking lines onto metal that won't rub off – Scriber. Marking lines parallel to the edge of metal – Odd-leg callipers. Marking lines at right angles to the edge of metal – Engineers square. Measuring/transferring the distance between two points or marking an arc or circle onto metal – Spring dividers. Mark the centre of a hole prior to drilling – Centre punch.

Page 98: Cutting and Shaping

Task 1

Removing material for a mortise – Mortise chisel (Wood). Cutting intricate curves and shapes – Coping saw (Thin material – plywood, acrylic, wood). Cutting straight lines, often across the grain of material – Tenon saw (Wood). Making a rip cut, parallel with the grain of material – Rip saw (Wood). Cutting larger scale materials – can handle curves and straight lines – Jig saw (Wood). Carving or cutting – ideal for removing thin pieces of material at a time – Bevel-edged chisel (Wood). Ideal for cutting straight lines – has replaceable blades – Hacksaw. Cutting sheet material – Tin snips/snips/shears (Metal). Cutting fine amounts of material away, in straight lines, internal curves and external curves – Half-round file (Wood, metal and edge of plastic). Shaping and smoothing the end grain of the material – Band facer (Wood).

Page 99: Clock Case Study

Task 2

1. Centre lathe.
2. Facing off.
3. **a)** Mark the aluminium shape by creating a template, then using a scriber to mark the metal. The waste could be removed using a guillotine or tin snips. The part will then need to be filed to create the desired curved shape.

b) Create a template and use a marker to transfer the shape onto the plastic's protective cover. Use an abra file or coping saw to remove most of the waste material. A file can then be used to smooth the edges of the plastic to get the correct shape.

4. Wood stain/dye or paint.

Page 100: Finishing Metal

Task 1

1. Filing – removes burrs and sharp edges.
2. Emery cloth – removes any remaining roughness after filing.
3. Wet and dry paper – creates a smooth finish.
4. Polish – Buffs the metal to a high shine.

Page 101: Finishing Plastic

Task 2

1. Cross file – This should remove all of the saw marks. It will leave scratch lines across the material.
2. Draw file – This should remove the scratches left from cross filing. It will leave smaller marks along the edge of the plastic.
3. Wet and dry – This should remove the marks from draw filing.
4. Polish – This should bring up a smooth and shiny finish.

Task 3

Wood

Natural/transparent – vegetable oil, danish oil, lacquer, varnish
Coloured – stain (allows grain to be visible), paint, spray paint, creosote (for outdoors)
Prepare for finish – remove pencil marks and lines using P60 grit sand/glass paper, rub down to remove smaller imperfections (P120 grit paper), raise the grain (dampen the wood and use wet and dry paper), repeat wet and dry until finish is smooth, ready to apply finish.

Metal

Edges – file rough cut marks, emery cloth, polish
Sheet material – wire wool (removes scratch marks), soap and wire wool (brings up shine)
Coloured finish – spray paint, dip coat, paint

Plastic

Edges of acrylic (4 stage process) – 1. Cross file, 2. Draw file, 3. Wet and dry, 4. Polish

> **Link to finishes mind map**
> https://collins.co.uk/pages/scottish-curriculum-free-resources